neural mechanisms of

COLOR VISION

Double-Opponent Cells in the Visual Cortex

neural mechanisms of

COLOR VISION

Double-Opponent Cells in the Visual Cortex

Bevil R. Conway, Ph.D.

Department of Neurobiology
Harvard Medical School

KLUWER ACADEMIC PUBLISHERS
Boston / Dordrecht / London

Distributors for North, Central and South America:
Kluwer Academic Publishers
101 Philip Drive
Assinippi Park
Norwell, Massachusetts 02061 USA
Telephone (781) 871-6600
Fax (781) 681-9045
E-Mail: kluwer@wkap.com

Distributors for all other countries:
Kluwer Academic Publishers Group
Post Office Box 322
3300 AH Dordrecht, THE NETHERLANDS
Telephone 31 786 576 000
Fax 31 786 576 474
E-Mail: services@wkap.nl

QP
483
C66
2002

 Electronic Services <http://www.wkap.nl>

Library of Congress Cataloging-in-Publication Data

A C.I.P. Catalogue record for this book is available
from the Library of Congress.

Printed on acid-free paper.

Printed in the United States of America.

The Publisher offers discounts on this book for course use and bulk purchases.
For further information, send email to mimi.breed@wkap.com.

Contents

Foreword

In 1967 Nigel Daw discovered a new type of cell in goldfish retina, which became termed "double opponent". The receptive fields of these cells were organized in center-surround fashion, but were more complicated than the center-surround fields described in the early 50s by Stephen Kuffler. Double-opponent cells were activated by small spots at one set of wavelengths and suppressed by light of a different set of wavelengths; the surrounds of these fields were also color coded: a cell might for example have a red-on, green-off center and a red-off green-on surround. Daw, who had trained in Edwin Land's laboratories, immediately realized the implications for human color vision. Double-opponent cells went a long way toward accounting for color contrast and for the phenomenon of color constancy and the related spatial-color effects Land had described. (In 1966 Torsten Wiesel and I had found color coded center-surround cells in the macaque lateral geniculate body, but these were single-opponent — for example, red on-center/ green off surround — and were wired just the wrong way around to explain spatial color contrast or color constancy. What part these cells play in primate color vision is still a mystery.) A few years later Nigel Daw and Alan Pearlman made a determined effort to find double-opponent cells in the monkey lateral geniculate, with no success. Meanwhile there were strong hints of the presence of double-opponent cells in monkey cortex.

In 1968, Wiesel and I had seen a few examples of cortical cells that we thought must be double opponent. Margaret Livingstone and I subsequently observed many more of these cells in monkey cortex, where they were concentrated in cytochrome oxidase blobs, but we had little success in obtaining explicit responses when we stimulated the surrounds by themselves, for example with annuli. To explain such difficulties Daniel Ts'o and Charles Gilbert proposed the concept of the "modified type 2 cell" in which light in the surround, regardless of wavelength, suppressed the responses from the center. The picture was further complicated by the failure of some authors to find cortical double opponent cells at all. Clearly more powerful techniques were needed. A clean definition of the borders of the center and surround required small-spot stimuli, but small spots gave impossibly weak responses. Bevil's answer was to bring out the small-spot responses by computer averaging. He did this by using eye position corrected reverse correlation, a technique that had been worked out a few years earlier by Margaret Livingstone. Here one stimulates with a series of spots at random positions and wavelengths, and asks, for each impulse, what were the positions and colors of the spot at some suitable time prior to the impulses. Next, the use of monochromatic light, as generated with interference filters, was necessarily messy given the overlap in the spectral sensitivities of the three cones. Cone isolation methods were obviously the answer. Typically one begins with a diffuse gray, on which a spot is generated in such a way that the activity of only one of the three cone types is affected. For most cells the intensities of light available with video monitors were too feeble to evoke detectable responses in cortical cells, even with computer averaging. By using colored backgrounds rather than gray, and by ignoring advice from various helpful colleagues, Bevil was able to achieve much more powerful cone-isolation stimuli. Computer averaging combined with cone-isolation led him to the first clear maps of double opponent receptive fields. These cells have relatively large receptive-field centers, so that the visual fields can be covered by a relatively small population. The problem of finding such rare cells was overcome by simple screening tests.

We now have the first clear demonstration of double opponent cells in the primate visual system. Given the temperament of those who work in the field of color vision there seems little doubt that heated debates will continue, but for the present, at least, the subject seems to be as close to settled as such things can be in science.

—David H. Hubel, M.D.
August, 2001

Abstract

How the brain represents color remains one of the most controversial topics in neurophysiology. We know that color is not represented by merely mixing the three classes of cones (L, M and S, or loosely red, green and blue) as a painter mixes pigments, but rather through an opponent mechanism. This is demonstrated by the fact that some colors are exclusive of others; specifically red is exclusive of green (we cannot even conceive of reddish greens), and blue is exclusive of yellow. Yet how these antagonistic chromatic axes are represented in the cortex has been a mystery. I investigated this question by mapping the spatial and temporal structure of the cone inputs to single neurons in the primary visual cortex of the alert macaque. Macaque monkeys are a useful model for human color vision because their performance on visual tasks is virtually identical to that of humans.

Color cells had receptive fields that were often Double-Opponent, an organization of spatial and chromatic opponency sufficient to form the basis for color constancy and spatial color contrast. The ratio of L and M cone input into red-green cells was roughly the same for all cells, suggesting that this population represents a single chromatic axis (color cells were screened for L vs. M opponency so very few blue-yellow cells were studied). Most color cells responded to stimuli of all orientations and had circularly symmetric receptive fields; some showed a coarse orientation preference that was reflected in the receptive fields as oriented Double-Opponent subregions. Red-green cells often responded to S-cone stimuli. Almost all color cells gave a bigger response to a color if it was preceded by an opposite color, suggesting that these cells also encode temporal color contrast. In sum, color perception is likely subserved by a subset of specialized neurons in primary visual cortex. These cells are distinct from the sharply orientation-tuned, and often direction-selective cells that likely underlie form and motion perception. Color cells establish three color axes (red-green, blue-yellow and black-white) that are sufficient to describe all colors; moreover these cells are capable of computing spatial and temporal color contrast — and probably contribute to color constancy computations — because the receptive fields of these cells show spatial and temporal chromatic opponency.

Preface

Thinking about the nature of light and visual perception seems to be an activity to which humans are particularly prone — if the disproportionate number of distinguished minds that have devoted themselves to the topic is any measure (Helmholtz, Hering, and Hubel being the three H's that come to mind). Perhaps it is the combination of a well-developed visual system and a particularly large chunk of cortex with which to consider it that predisposes man to this pursuit. Needless to say, I too am drawn to the puzzle of how we see.

With the philosophers came the appreciation that visual perception is distinct from the physical cues on which it is based. Light enters the eye and is focused by the cornea and lens on the retina, but our brains do not develop the image like photographic film. Rather visual perception is a reconstruction of biologically relevant aspects of the physical world.

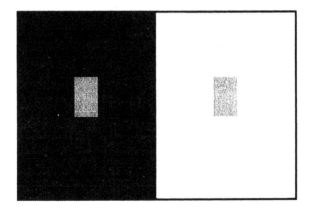

Figure P.1. The brightness of a gray square changes depending on the context.

Often these aspects are relationships between parts of the retinal image, for example how bright one part is relative to another, as distinct from the parts themselves (Figure P.1).

Artists learn this implicitly: they paint a bright object not by using white, but by using white next to black. This trick, called *chiaroscuro*, enables artists to generate the illusion of luminous objects despite the physical limitation that the reflected light of a painted surface can never be as luminant as light sources in the real world, which often they are trying to depict (Rembrant comes to mind). The appreciation that the visual system is not a camera is one that must be achieved through independent observation. Indeed a child need never become aware that he has a hole in his visual field the size of a quarter held at arm's length, 18 degrees from his center of gaze. Evolution has equipped him not only with the tools with which to acquire information about the world but also with the ignorance about the shortfalls of the tools. It is this robust ignorance that makes the blind spot demonstration so surprising and delightfully ironic every time.

Psychophysicists and artists have accumulated enormous catalogues of visual illusions, the relative brightness illusion in Figure P.1 and the blind spot illusion being just two of them. But every illusion elicits from the viewer a similar perplexed response (e.g. "Well how bright is it really?") as if these illusions represent the one circumstance in which vision has "failed" us. Of course the moral is that the illusions do not illuminate failure, but rather success. They provide insight into what the visual system is constantly (and usually very effectively) doing: extracting biologically relevant pieces of information from the physical world and reducing them to meaningful attributes. For example, it is unlikely that knowledge of the absolute luminance of gray rectangles (as in Figure P1) would ever be of advantage to survival, in the way that knowledge of the relative brightness could be. A frog would find a fly nutritious on a cloudy day or a sunny day. Under each condition, the *relative* brightness of the fly against its background is not that different even though the absolute amount of light reflected from the fly is different.

A naive understanding of the dependence of visual perception on physical cues could lead us to conclude that the gray rectangle in Figure P.1 is reflecting a different amount of light in the two situations — because its brightness differs in the two situations. But we must be careful. A nice example of this was provided by Hering (Hering, 1964, pg. 14-15) when he considered the absolute amount of light reflected from black text on a white page under different lighting conditions. He noted that the black of the black text observed at midday (in bright daylight) is about three times as intense as the white of the page on which it sits, observed in the early morning (in dim daylight). Yet we report the letters of the text as black and the color of the page as white in both situations. Our visual systems have successfully reduced a physical cue (the amount of light) into a relevant attribute: the *relative* brightness of the letters; we report this attribute as a property of the text (it *is* black).

One attribute of the world to which we are particularly sensitive is color. It is not hard to see why evolution has equipped us with this sensitivity: colors enable primates to discriminate suitable food (ripe fruit) and sexual partners (females in estrus/ males in plumage). In this book, besides attempting to understand the broad question of the neural basis for color vision, one of my main motivations is a desire to understand how it is that the visual system represents color based on physical cues which do not always directly correlate with our perception.

Colors are not simply cues for discriminating objects; they also have emotional significance. It is perhaps this quality of color that has driven color vision research and may explain the passionate controversies that fill this field's history. It is my sincere hope that this book does not exacerbate the tensions in the field but rather serves as a contribution to an empiric basis on which to resolve them.

— August, 2001

Acknowledgements

I am grateful to Margaret Livingstone and David Hubel for their mentorship and friendship. Without them, I could not have done this work. I am indebted to Mimi Thompson Breed, my editor at Kluwer, for her support and comments on the manuscript. I thank Doris Tsao for many useful discussions; David Freeman for all the computer programming; Mike Lafratta, for machining; and Tamara Chuprina for animal care. I thank John Assad, Richard Born, Denis Baylor, Arthur Bradley, Andrew Stockman, Nigel Daw, Kevin Duffy, Richard Masland, Carla Shatz, Jonathan Trinidad and especially Clay Reid for their useful comments. This work was funded by the Natural Sciences and Engineering Research Council of Canada, the Mind Brain and Behavior Initiative (Harvard) and the National Institutes of Health grant EY 10203 (to Margaret Livingstone) and EY 00605 (to David Hubel).

Chapter 1
A history of color research

"To explain colors, I suppose, that as bodies of various sizes, densities, or sensations, do by percussion or other action excite sounds of various tones, and consequently vibrations in the air of different bigness; so the rays of light, by impinging on the stiff refracting superficies, excite vibrations in the ether,—of various bigness; the biggest, strongest, or most potent rays, the largest vibrations; and others shorter, according to their bigness, strength, or power: and therefore the ends of the capillamenta of the optic nerve, which pave or face the retina, being such refracting superficies, when the rays impinge upon them, they must there excite these vibrations, which vibrations (like those of sound in a trunk or trumpet) will run along the aqueous pores or crystalline pith of the capillamenta, through the optic nerves, into the sensorium;—and there, I suppose, affect the sense with various colors, according to their bigness and mixture; the biggest with the strongest colors, reds and yellow; the least with the weakest, blues and violets; the middle with green; and a confusion of all with white, much after the manner that, in the sense of hearing, nature makes use of aerial vibrations of several bignesses, to generate sounds of divers tones; for the analogy of nature is to be observed."
— Sir Isaac Newton, 1675 (in Newton, 1978)

1.1 Introduction

A world devoid of color is a gray and bleak one. Despite this, the specific benefits of color vision are difficult to quantify. Picasso said flippantly "when I run out of blue I use red" by which he meant that it is the brightness (or value) of a pigment and not its color that describes the 3-dimensional shape of objects. Matisse demonstrated this point beautifully in his painting *La Femme au Chapeau* (*The Woman in the Hat*, Paris, 1905; San Francisco Museum of Modern Art). A grayscale version shows that the values of the pigments preserve the shape of the woman's face quite well: the color transitions do not interfere with an accurate representation of the play of light across her face (Livingstone, 2002). That the painting reads well as a face despite the radical color transitions shows that color is not an important cue to shape. Indeed, object shapes are easily recognizable even in dim light when color vision is absent. Moreover, many people function perfectly well with impaired color perception: as many as one in twelve men are red-green color-blind and many of them are unaware of it. But color cues are useful. In monkeys, for example, they assist the discrimination of nutritional foods (e.g. red berries) and of suitable procreative partners (e.g. male birds with the most colorful feathers). In humans, color is more than a cue for discriminating objects. Unlike other object attributes like shape and texture, colors have emotional significance. One is green with envy, red with anger, blue with sadness. Indeed, Matisse certainly was aware of the impact of his surprising choices of color in his painting *La Femme au Chapeau*: they push the portrait past mere representation by showing Matisse's emotional response to the subject. It is perhaps this emotional quality of color that has fueled color vision research among philosophers, physicists, psychologists, physiologists and artists; and it may also explain the passionate controversies that fill this field's history.

1.2 Color vision: what is it?

Color perception is our ability to use differences in spectral composition of light to discriminate objects, surfaces or light sources. The color we assign an object depends not only on the particular spectral composition of light reflected from it but also on the light reflected from surrounding objects. Color vision has evolved in several species of animals, such as insects and birds; color vision is crude in most mammals except some primates. Here I will focus on the neural mechanism of color vision in Old World monkeys, specifically the macaques, because behavioral, anatomical and physiological studies indicate that their color vision is virtually identical to that in man (DeValois et al, 1974; Sandell et al, 1979).

1.3 Trichromacy

We are indebted to Newton for pointing out that white light comprises separable parts, each having a different wavelength, and that a given visual effect may be brought about using different combinations of wavelengths (Newton, 1671). For example, an identical perception of yellow can be produced by monochromatic yellow or by a combination of two monochromatic lights, one that appears red when used alone and one that appears green. But "for the trained physicist, it is easy for the idea of different wavelengths to become so firmly associated with the different hues of the spectrum that he finally ascribes a specific hue to each individual kind of light ray, and names the rays by the hues rather than according to their frequency or wavelength." (Hering, 1964, pg. 289). The misperception that it is the wavelengths themselves that are colored is entrenched in our education — and not without reason. Under most circumstances it is true that short-wavelength light appears blue and long-wavelength light red. But, as Edwin Land showed (see below) with his powerful demonstrations, it is not always the case that wavelength accurately predicts perceived color. In the same way the amount of energy reflected off the gray square in Figure P.1 (see Preface) is not directly correlated with its brightness, the spectral distribution reflected off an object is not directly correlated with its color. Thus "the full appreciation of natural phenomena, as we see them, must go beyond physics in the usual sense" (Richard Feynman).

In 1802, Thomas Young asserted that colors come about because of a property of man's visual system and not a property of the wavelengths themselves (Young, 1802). Young was probably inspired by the work of George Palmer, an English glass maker in the mid 18th century (Mollon, 1993), to postulate to the Royal Society of London that "there are in the eye three kinds of nerve-fibres, the first of which, when irritated in any way, produces the sensation of red, the second the sensation of green, and the third that of violet." (Young, paraphrased by Helmholtz, 1873, pg. 244-255). Helmholtz popularized this idea of trichromacy (Figure 1.1); indeed if it had not been for the eloquent championing by Helmholtz, Thomas Young might have been known only as the man who deciphered Egyptian hieroglyphics.

The trichromatic theory was very powerful for it resolved one of the most profound problems in color. "As it is almost impossible to conceive each sensitive point of the retina to contain an infinite number of particles, each capable of vibrating in perfect unison with every possible undulation, it becomes necessary to suppose the number limited, for instance, to the three principal colors, red, yellow, and blue, of which the

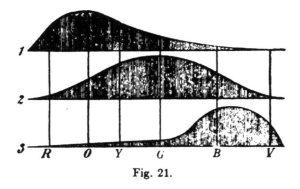

Fig. 21.

Figure 1.1 Young and Helmholtz proposed color was subserved by three kinds of "nerve fibres", with these absorption curves (Helmholtz 1909, his Fig. 21) .

undulations are related in magnitude nearly as the numbers 8, 7, and 6; and that each of the particles is capable of being put in motion less or more forcibly, by undulations differing less or more from a perfect unison; for instance the undulations of green light being nearly in the ratio of 6 ½, will affect equally the particles in unison with yellow and blue, and produce the same effect as a light composed of those two species: and each sensitive filament of the nerve may consist of three portions, one for each principal color" (Young, 1802).

We now know that color perception begins in the retina with three classes of photoreceptors, the cones. Psychophysical studies based on color matching (Stiles, 1939; reviewed by Lennie and D'Zmura, 1988), reflection densitometry [which involves shining lights into the living human eye and measuring the amount reflected back (Rushton, 1958, 1962)], and studies of color vision in so-called colorblind individuals (Rushton, 1975) supported the notion that three classes of photoreceptor were responsible for color vision. These three classes were interpreted as the physiological basis for the fact that almost any color could be produced from only three suitably chosen primaries (Maxwell 1856), a fact with which artists have almost always been well acquainted.

But it was with the development of a technique that enabled responses of single cones to be measured (microspectrophotometry!) that distinct red and green cones (Brown and Wald, 1963) and then blue cones (Marks et al., 1964) were identified; this has been confirmed in both humans and old-world monkeys (Dartnell et al, 1983; Bowmaker et al, 1980). Cones are more densely packed in the portion of the retina corresponding to the center of gaze (the fovea), and become less dense in the periphery. A second group of photoreceptors, the rods, have a different distribution: they are actually absent from the fovea and become prominent in the near-periphery (Polyak 1957, pg. 270-4). Rods function best in dim light, when the cones do not function well (Rodieck 1998, pg. 180). Because we only have one class of rods, and a comparison between at least two classes of photoreceptor is required for color vision, rods for the most part are not involved in color perception (which explains why we do not see color in dim light).

1.4 The cones

The three classes of retinal photoreceptors which mediate color vision are called the cones; each class of these specialized cells contains a unique photopigment: the S cones contain a photopigment that absorbs shorter wavelengths optimally (peak 440nm); the M cones, one that absorbs middle wavelengths best (peak 535 nm); and the L cones, one that absorbs middle and long wavelengths best (peak 565 nm) (Figure 1.2).

Though each class of cones has its own peak sensitivity, each is somewhat sensitive to wavelengths throughout most of the spectrum. This means that a single cone (or class of cones) is color-blind because it cannot distinguish between a dim light of optimal wavelength and an intense light of less optimal wavelength. This is known as the principle of univariance (Brindley, 1957; Rushton, 1972). At any given point in the retina there is only one cone so the retina is actually colorblind on a spatial scale of single cones. The cones are often referred to by color names (blue, green and red). But this is misleading not only because a single cone class is colorblind but also because blue, green and red are not the color names that we assign to the region of the spectrum to which each class is maximally sensitive. This is especially true of L and M cones whose peaks, only separated by 30 nm, are in the yellow.

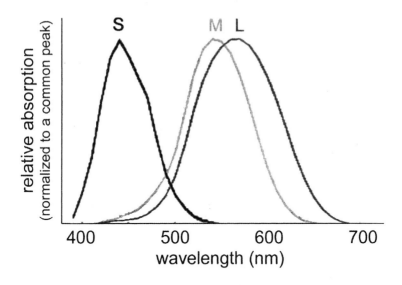

Figure 1.2. The cone absorption spectra (normalized to a common peak (adapted from Wandell, 1995).

It was once taken for granted that the cones of the primate retina would be regularly distributed (to facilitate uniform sampling of wavelength), as is the case in the goldfish retina. Though the S cones are regularly distributed [(except in the center of the fovea where they are absent (Curcio et al, 1991)], the L and M cones are surprisingly patchy and possibly random (Roorda and Williams, 1999; Figure 1.3). The resulting clumpiness facilitates our ability to detect fine-grained luminance variations: neighbouring cones having the same peak sensitivity are better able to discriminate subtle spatial luminance fluctuations than neighboring cones with different peak sensitivities. But this clumpiness is at the cost of color resolution. This is evident perceptually: our visual systems have a lower resolution for objects that differ only in color (rather than in both color and luminance). Two objects that differ only in color are called isoluminant (such colors will come out roughly the same gray in a grayscale copy of them).

8

You can often find examples of isoluminant colors in famous paintings and in advertising (because the visual effect is eye catching, see Livingstone and Hubel, 1988 and Livingstone 2002). Curiously the question of how wavelengths are sampled evenly across the retina remains unresolved. It may be the case that the clumps of M cones, which reside in a sea of L cones, reside directly next to S cones (Figure 1.3; Conway, 2000; Conway, 2001), which are regularly spaced; thus, though the individual cones appear randomly distributed, on a coarse scale the clumps of identical cones are not random.

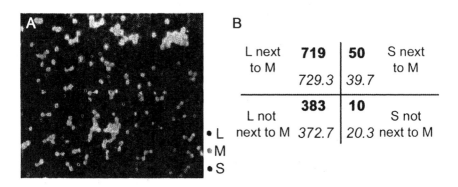

Figure 1.3. S cones are adjacent to M cones more often than expected by chance, as determined using the figures from Roorda and Williams (1999) *Nature* **397**, 520-2. (**A**) is adapted from Roorda and Williams (1999) and shows the cone mosaic in the living human eye. (**B**) is a chi square comparison of the S and L cone distribution with respect to the M cones. All the cones from Figure 3 of Roorda and Williams were counted. The expected values are shown in italics; the observed values are shown in bold. X^2 value = 8.34 indicating that S cones reside next to M cones more often than expected by chance (p<0.005).

A second striking feature of the cone mosaic is that the very center of the fovea (corresponding to about a tenth of a degree of visual angle) is devoid of S cones. It so happens that our eyes are focused for ~550 nm light, where the L and M cones have their peak sensitivities. Consequently shorter wavelength light will be blurred. Evolution may have selected against having S cones in the center of the fovea where high spatial acuity is the goal because the short-wavelength light to which the S cones are most responsive will be an unreliable source of spatial information. The absence of S cones in the center of the fovea has surprisingly little impact on color vision; this is probably because the spatial extent of the S cone hole is finer than the coarse resolution of color vision.

The wavelength sensitivity of a given cone is attributed to the specific photopigment that it expresses (Nathans, 1999). The genes for the M and L photopigments are both encoded on the X chromosome and are fairly similar in sequence. The similarity in sequence suggests that the M and L photopigments are derived from a common ancestral gene that duplicated not so long ago — around 30–40 million years ago, shortly after the geologic split between Africa and South America. The similarity of the M and L gene sequences predisposes them to recombination during meiosis. These recombination events have led to a polymorphism of the L and M photopigments, which is more common in males because males only have one X-chromosome on which to rely for their M and L photopigments. The polymorphism can be manifested as a complete loss of L cones (protanopia), loss of M cones (deuteranopia) or, more frequently, the expression of a mutant M/L hybrid. It is these polymorphisms that underlie the spectrum of red-green color blindness, the most famous case of which is that of Sir John Dalton (a deuteranope), who in 1794 was the first to describe the condition (Daltonism is another word for color-blindness). The deletion of the S cone gene, which resides on the seventh chromosome, is possible (tritanopia), but rare because it requires a deletion of both copies of the S gene. Color matching experiments in dichromatics (people who express only two photopigments) have been very useful in establishing the spectral sensitivities of cones (Vos and Walraven, 1971; Smith and Pokorny, 1975; Stockman and Sharpe, 2000).

1.5 Opponency

The power of the trichromatic theory of color held sway until 1880 when the German psychologist Ewald Hering offered an alternative theory. Hering observed that we are not capable of perceiving a continuous mixture of colors as would be predicted by the trichromatic theory. Instead some colors seem exclusive of others; for example we cannot perceive (or even conceive of!) reddish-greens or bluish-yellows. Hering argued that color was determined not by three classes of receptor but rather by the activity of three opponent mechanisms: a red-green mechanism, a blue-yellow mechanism and an achromatic black-white mechanism. The battle between Hering and Helmholtz and their supporters is the source of much animosity in the field of color, still today, even though the development of modern physiological and anatomical techniques has reconciled them (see below and Judd, 1951, page 834).

It is interesting to reflect on the controversy between Hering and Helmholtz, especially because some of it (I believe) had its roots in a miscommunication and not in a difference of opinion. I do not think Helmholtz could have believed in the strict interpretation of the trichromatic theory (that one "nervous fibre" stimulated in isolation produces a predictable color sensation). He states "there is at the furthest limit of the visible field a narrow zone in which all distinction of colors ceases and there only remain differences of brightness. In this outermost circle everything appears white, gray or black. Probably those nervous fibres which convey impressions of green light are alone in this part of the retina" (Helmholtz 1873, pg. 248). It is unclear to me what narrow colorless zone Helmholtz is referring to (I haven't noticed it), but he states explicitly that the percept resulting from the stimulation of a single cone class is colorless.

While the psychologists were waging war on each other, the neurobiologists were developing better and better recording techniques. Adrian and Mathews (1926) developed an approach using the eel to record the simultaneous activity of large numbers of fibers in the optic nerve while they stimulated the retina with light spots of various sizes and intensities. They found that smaller and less intense stimuli resulted in smaller responses. They also showed that an increase in the size of a stimulus could compensate for a decrease in its intensity. They used this to draw a parallel with the psychophysical observation in people that a less intense spot can be made as salient as a more intense one if its size is increased a bit. These studies set the stage for *single* unit recordings in the frog (Hartline, 1939), cat (Kuffler, 1953) and fish (Svaetichin and MacNichol, 1958).

It soon became clear that the question with which the psychologists were wrestling — is it the three-color system of Young-Helmholtz or the opponent-colors system of Hering? — might be resolved by recording the responses of single units in the visual system while stimulating the retina with different wavelengths of light.

In a wonderful paper, Svaetichin and MacNichol (1958) show the spectral tuning curves for various cells of the fish retina (Figure 1.4). Though Svaetichin (1956) originally thought his recordings represented receptor potentials, Tomita (1957) and MacNichol and Svaetichin (1958) subsequently showed that the responses arose more proximally, probably in horizontal cells (MacNichol, 1964). The confusion is understandable because at that time methods for recording and labelling single cells had not been developed; recording electrodes were placed in the retina blindly, according to depth. Remarkably, some of the cells showed opponent responses to different wavelengths (Figure 1.4).

In fact, in the species of shallow-water-dwelling fish from which they were recording they found not only red-green cells but also blue-yellow cells. Thus it was concluded that the "psychophysical observations [of opponent colors in humans] are easily explained on the basis of a subtraction of two different opponent receptor responses" (Svaetichin and MacNichol, 1958). Of course we have no idea whether or not these fish see in color (though psychophysical studies in a different species of fish, the common goldfish, have since been done; Ingle, 1985). This uncertainty directed De Valois and his colleagues to look for color opponency in an animal with color vision closer to our own, the macaque monkey (De Valois et al., 1958). Recording single units extracellularly in the lateral geniculate nucleus (LGN), De Valois et al. (1958) were able to find color opponent responses similar to those found by Svaetichin and MacNichol in the fish retina.

In sum, both the trichromatic theory and the opponent theory are correct. In fact Müller (1930) provided a framework in which these two theories could co-exist and explain most phenomena of color vision (Judd, 1949, 1951), but it was not until the development of precise recording and anatomical techniques (e.g. Dacey and Lee, 1994; Calkins, 1998) that Muller's "zone theory" (also known as stage theory) was given a firm biological basis. The three classes of photoreceptors (L, M and S cones) represent the trichromatic theory; the retinal ganglion cells (and lateral geniculate nucleus cells) that receive opponent inputs between cone classes and which are positioned at a higher stage (or zone) of visual processing, represent the opponent theory. Indeed this reconciliation was appreciated by Hering, who acknowledged that "the Young-Helmholtz three-color theory 'could, with some modifications,

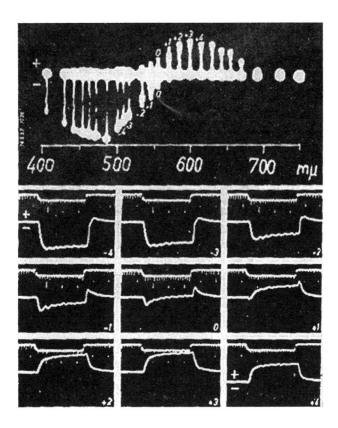

Figure 1.4. Intracellular recording from the retina of the *Mugilidae sp.* showing chromatic opponency. (Svaetichin and MacNichol, 1958). The bottom nine panels show enlargements of the responses used to construct the spectral tuning curve (top). In each panel, the top trace shows the stimulus duration (time is from left to right); the bottom trace shows the voltage displacement of the cell: suppression (-) is deviations below the horizontal, excitation (+), above. Note the tiny numbers in the bottom right corner of each of the nine panels and the corresponding numbers along the envelope of the spectral tuning curve — these indicate the wavelength used in each of the nine panels (e.g. the trace in the upper left panel shows the response when the cell was stimulated with 510 μm light).

very well exist side by side [with the Hering opponent-color theory] if one strictly distinguished between the *process of excitation* and the *process of sensation*,' and used the three-color theory for the former and my theory for the latter" (Hering quoting Aubert (1865), (Hering, 1964, pg. 48). Moreover, given the cone absorption spectra (Figure 1.2) it is easy to see why color must be an opponent process: the L and M cone fundamentals are so similar that the only way we could perceive long-wavelength light as distinct from middle wavelength light (i.e. red and not yellow) would be by subtracting the M fundamental from the L fundamental.

1.6 Spatial structure of receptive fields

While everyone seemed to be occupied with the neural basis for Hering's opponency, a Hungarian, a Canadian, and a Swede working at Johns Hopkins were plugging away trying to figure out how the visual system encodes spatial information (Kuffler, 1953; Hubel and Wiesel, 1959). Instead of using full-field stimuli, as was commonly done by other groups (De Valois et al., 1958), these scientists were using small spots. Their rationale was simple: in order to figure out how the visual system encodes spatial information one has to use spatially restricted stimuli. Indeed such stimuli enabled Hartline to discover lateral inhibition in the eye of the horseshoe crab (*Limulus polyphemus*) (Hartline, 1949, 1956) for which he won the Nobel prize.

Wiesel and Hubel's (1966) surprising discovery was that most neurons in the primate lateral geniculate nucleus were not only chromatically opponent (as De Valois and colleagues (1958) had shown) but that they were also spatially opponent (Figure 1.5).

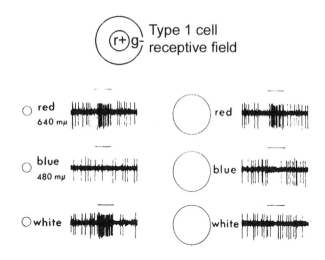

Figure 1.5. Spatial and chromatic opponency of a Type I cell, from the monkey lateral geniculate nucleus (Wiesel and Hubel, 1966).

Wiesel and Hubel distinguished "Type I" cells from "Type II" and "Type III" cells. The receptive field of a Type II cell consists of a single region in which different wavelengths produce different responses (excitation or suppression). For a given eccentricity, Type II cells have considerably larger receptive fields than the receptive field centers of Type I cells, and are much less numerous (Wiesel and Hubel found that only 7% of the neurons in the lateral geniculate nucleus were Type II; 77% of the neurons were Type I). Type III neurons form a catch-bag for the rest of the cells in the lateral geniculate; these cells show no chromatic opponency, though their receptive fields are subdivided into center and concentric, antagonistic surround (with some centers inhibitory and others excitatory). The spatial and chromatic opponency of Type I cells was a remarkable finding; and, on first inspection, this spatial opponency *seemed* to be the solution to the latest dilemma to occupy color psychophysicists: the color constancy dilemma.

1.7 The color constancy dilemma

In 1959 Edwin Land (inventor of instant photography) was exploring the nature of light using cameras, color filters and slide projectors (Land, 1959a,b). Land had become fixated on the problem of color constancy and the surprising extent to which the visual system is color constant.

The physical cues that the visual system uses to construct a color percept are the different spectral distributions (i.e. wavelengths) reflected from objects. The spectral distribution of light that reaches the eye is a product of the absorptive properties of the object's surface and the spectral properties of the illuminant. In normal daily life the spectral properties of the illuminant change. For example, the illuminant of a bright sunny day, under a blue sky, is rich in short wavelengths, whereas the light of a tungsten light bulb is rich in longer wavelengths. In fact at any given time the illumination conditions will vary across a given scene. The paradox of color vision is this: despite these different illumination conditions, and the resulting difference in light reflected from a given object, the color we assign to the object is to a large degree illuminant-independent. A red apple is red, for example. It is not that a red apple is only red when viewed under a certain illumination condition like a blue sky. Just as brightness is not equated to the intensity of light entering the eye (see Preface, Figure P.1), color is not simply equated to wavelength. This is familiar to color photographers who are constantly plagued by the effects of different illumination conditions: a scene photographed under tungsten light comes out with a reddish cast and a scene outside on a sunny day with a bluish cast. This is in contrast to our perceptions of the scene under the two illumination conditions: we see neither a reddish cast nor a bluish cast.

It is easy to see why our visual systems have evolved this way. If we were to assign a color to an object based solely on the light reflected from it, then we would assign different colors to the same object depending on the conditions under which the object was viewed. That the color assigned to given objects varies so little is a property of color vision known as color constancy, and the benefit it confers upon us is easy to appreciate: colors become properties of objects (which are constant) and not viewing conditions (which are continually changing).

1.8 Edwin Land's Retinex theory

Edwin Land did not discover that the visual system was color constant — though his demonstrations certainly convinced any non-believers (Land, 1977). What he did do was use his demonstrations to take a computational approach to color vision; this approach produced a testable hypothesis for how the visual system achieves color constancy. A Land demonstration began with a large picture constructed from irregular patches of different colors — he called such pictures Mondrians because he thought they bore some resemblance to the paintings of Piet Mondrian (I don't think they do: Mondrian was concerned with properties of grids, not irregular patches of color). Land illuminated the Mondrian with 3 projectors, one emitting short-wavelength light, another medium-wavelength light and a third long-wavelength light. When all three projectors were used in an appropriate proportion the combined light appeared white. Using a spectrophotometer, Land determined the amount and wavelength composition of the light (i.e. the spectral distribution) reaching the eye from a given patch in the Mondrian. He then chose a different area in the Mondrian, directed his spectrophotometer to that area, and altered the relative intensity of each of the three illuminating projectors until the spectral distribution coming from that area was the same as that from the first area. Thus in two differently illuminated conditions, the spectral distribution coming from two different areas was the same. The shocking thing is that the appearance of the two areas is not the same! In fact, as Land showed, the color name we assign to the areas is largely independent of the two illumination conditions. This is surprising because one might have expected that the only information on which the visual system can base a color judgment would be the spectral distribution reaching the eye. Our eyes presumably have little information about the spectral content of the illuminant and we certainly have no *a priori* information about the absorptive properties of the surface. So the puzzle remained: how does the visual system achieve a different color judgment for the two areas if the amount and wavelength composition of light coming from the two areas is the same?

As impressive as the Land demonstration was, it simply reiterated what psychologists had long known: under different illumination conditions the colors of objects do not appear to change. Helmholtz discussed the constancy of colors at length, as if the phenomenon were well known (Helmholtz, 1909, pg. 172-86). In fact as early as 1672 it was appreciated that different natural illuminants (e.g. blue sky versus candlelight) had different wavelength compositions, but that the color of objects viewed under these different conditions changed very little (Otto von Guericke, as discussed by Land, 1977). The different spectral distributions of different illuminants was also demonstrated by Ragona Scina (Hurvich, 1981, pg. 153) and Benjamin Thomson (Thomson, 1794), who showed that the two shadows of a single object cast simultaneously by two different illuminants were different colors (Brown, 1970, pg. 69).

Edwin Land's contribution came with his invention of an algorithm, Retinex, which predicts the color an observer would assign to an area given the ensemble of values for the absolute energy reaching the eye from different regions of the image. His algorithm depends on a spatial comparison of the amount of energy coming from different areas across an image (Land, 1977), and makes a prediction that the human visual system makes a similar spatial comparison.

The algorithm first requires that the relative cone activities produced by the light reflected from every region of the image (not only the area for which you wish to determine the color) be determined. Land derived these relative cone activities with "the help of filters and appropriate film emulsions" that "isolate" the "lightnesses of the short-, middle- and long-wave sets of receptors". These "retinex records" "simulate the response of the three cone pigments", for example "strawberries and radishes are light on the long-wave record, darker on the middle-wave record and darkest on the short-wave record". Land and his colleagues determined the relative cone activities (which he called "integrated radiances") "by using a photomultiplier in conjunction with a version of the retinex filters". This is analogous to multiplying each of the cone absorption curves (see Figure 1.2) by the reflected spectral distribution. In this way, Retinex obtains a triplet of values for every region in the image, representing the integrated radiance at each point for the three cone classes.

In the second step, done separately for each cone channel, Retinex calculates the ratio of integrated radiances between adjacent regions. The ratio is subjected to a threshold test: the ratio of integrated radiance between two adjacent spots of an image is only considered significant if it deviates from unity by some defined amount (Land chose 0.003, Land 1977). This eliminates the impact produced by gradual changes in illumination across an image, and exaggerates the impact of abrupt changes. The threshold test is a critical part of Retinex because it accounts for an important part of visual perception: that gradual changes across an image are not well detected but abrupt changes (such as edges) are.

The whole image is processed in terms of the ratios at closely adjacent points, producing dimensionless numbers that are independent of illumination. Retinex then compares the ratios of adjacent points along many paths taken across the image, which results in a "sequential product" that approaches the ratio of reflectances between the first area and any given final area, regardless of the distribution of illumination. In this way Retinex produces a rank order of lightnesses across the image. The rank order for different areas in an image is obviously different for each cone channel: for example an area that would be called "red" would have a higher rank order (i.e. a higher lightness) for the L cone channel than for the M cone channel. The triplet of lightness values (one value for each of the cone channels) defines the illumination-independent color sensation at each point.

In summary, Retinex argues that color perception depends on three comparisons: first, local comparisons of the activity of a given cone class between adjacent areas (that eliminate the effect of gradual changes in illumination); longer-range comparisons of the activity of a given cone class across the entire image (that produce a rank order of cone activities across an image for each cone channel); and finally, a comparison of the relative activity of the three classes of cones at each point in the image (which produces the color sensation).

Retinex underscores the fact that the color we assign to a given surface is determined by the spatial chromatic context of the surface. One corollary of this is that our perception of a given pigment's color is shaped by the context in which it appears. This is well known by artists (Albers, 1963; Itten, 1966) and is used to advantage by advertisers who, with the intent of making a red area more salient, place it against a green background.   Of course Retinex is only an algorithm and makes no assertion about where or how the visual system makes this computation — and the name of the algorithm reflects this uncertainty (*retin*a or cort*ex*).   Land et al. (Land, 1983) subsequently did a rather clever experiment that showed that chromatic interactions between different colored regions across the vertical midline were compromised in a person with a cut corpus callosum (which dissociates the cortical processing of left and right visual fields), suggesting that the spatial comparison called for by Retinex is accomplished in the cortex.

1.9 The neural machinery for color constancy

The retinal ganglion cells and the lateral geniculate nucleus cells that are excited by one cone class but suppressed by a different cone class or classes (Type I and Type II cells, Figure 1.6A and B) clearly could subserve wavelength discrimination. But as the Land demonstration proved, color is not simply wavelength discrimination.   Rather, color is achieved through a *spatial* comparison of wavelengths across an image.   A cell having a receptive field fed by a single cone class in a spatially opponent fashion could be the building block for such a comparison — for example an L-on center, L-off surround cell.   Importantly, such a cell would go a long way in explaining color constancy. A uniform field of light would activate both subregions of such a cell's receptive field but would have opponent effects on the cell's firing and therefore would not be an effective stimulus.   Changing illumination conditions, which would involve a uniform stimulus to both subregions of the cell's receptive field, would therefore not drive the cell and thus not be perceived. Despite intensive searches, no retinal ganglion cells or lateral geniculate nucleus cells have been found to have this type of spatially opponent receptive field.

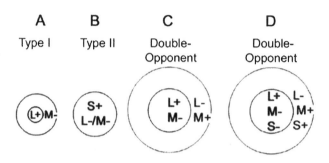

Figure 1.6. Diagrams of the receptive fields of color-responsive cells. A plus indicates excitation by the given cone, a minus, suppression. Type I and Type II cells are found in the lateral geniculate nucleus. Note that the surrounds of Type I cells infiltrate the centers, so that in this case (B), the center would also be suppressed by M cones. The sizes of Type II cell receptive fields are substantially larger than those of Type I cells. I found that the majority of *cortical* color cells with Type II-like centers are Double-Opponent (C), and many of them appear to receive S-cone input (D). The cortex also seems to contain a few Type I cells (see Figure 2.11).

As Wiesel and Hubel showed (1966), some lateral geniculate nucleus cone-opponent cells do have spatially opponent receptive fields (the Type 1 cells, Figure 1.6). One might have thought that these could represent the substrate for color contrast, but unfortunately the opponency is not sufficient because the center and surround are fed by different cone classes; moreover, the opponency between these classes is in the opposite direction than that required for color contrast.

So where in the primate visual system is such a comparison made? Nigel Daw showed in 1968 that some cells in goldfish retina have receptive fields that are both chromatically *and* spatially opponent and therefore capable of computing simultaneous color contrast (Daw, 1968). In principle such "Double-Opponent" cells could subserve color perception (and color constancy) in primates (Daw, 1968; Rubin and Richards, 1982; Livingstone and Hubel, 1984; Dufort and Lumsden, 1991; Courtney et al, 1995). A single Double-Opponent cell exceeds Retinex's requirement of a spatial comparison for one cone class: it is a spatial comparison for two cone classes. Moreover, independent component analysis of natural images predicts that color is coded by a separate color-double-opponent pathway, in conjunction with a luminance channel (Tailor et al, 2000).

Despite intensive efforts, Double-Opponent cells have not been found in the monkey retina or the lateral geniculate nucleus (Daw, 1972). Evidence consistent with Double-Opponent cells has been obtained in V-1 of anesthetized monkeys (Wiesel and Hubel, 1966; Poggio et al., 1975; Michael, 1978; Livingstone and Hubel, 1984), but this has since been interpreted as support for a formulation of these cells as "Modified Type II" (Ts'o and Gilbert, 1988). Modified Type II cells were defined as having a color-opponent center and a broadband surround that suppresses any effects of the center — these cells would seem to be incapable of solving color constancy. That Double-Opponent cells as originally described by Daw (1968) do not exist in primary visual cortex (Dow and Gouras, 1973; Vautin and Dow, 1985; Lennie et al., 1990) is the popular sentiment that is now in textbooks (Lennie and D'Zmura 1988; Lennie 1999, 2000).

The spatial structure of color cell receptive fields has typically been studied with simple colored spots and annuli (Michael, 1978; Livingstone and Hubel, 1984; Ts'o and Gilbert, 1988). The use of these stimuli makes arbitrary assumptions about the structure of the receptive fields that can be problematic: in at least one study, the annuli used to stimulate the surrounds likely encroached upon the centers (e.g. Figure 3 of Ts'o and Gilbert, 1988). Furthermore, these stimuli are not cone-isolating (any given stimulus changes the activity of all three cone classes), so the cone inputs can only be inferred. To describe the cone inputs, some have used DKL stimuli (Derrington et al., 1984; Lennie et al., 1990; Cottaris and De Valois, 1998). These stimuli modulate color between two states, simultaneously increasing the activity of one cone class while decreasing the activity of a second. Unfortunately, only full-field versions of these stimuli (uniform fields or gratings) were used, which would have confounded the responses of center and surround making it impossible to directly describe Double-Opponent cells. Moreover, DKL stimuli make it impossible to ascribe a cell's response to the increase in activity of one cone class as opposed to the decrease in activity of the second cone class, since both happen simultaneously.

Because the spatial structure of color cell receptive fields could clearly be important for subserving color constancy, and therefore color perception, my first experiment was to address the question directly (Chapter 2): what is the spatial structure of the cone inputs to cortical color cells? I did so using stimuli that selectively modulate the activity of a single cone class at a time ("cone-isolating" stimuli; Donner and Rushton, 1959; see Estevez and Spekreijse, 1982, for review). Of the cortical color cells with Type II-like centers, all the cells with significant surround responses were Double-Opponent.

From the receptive-field maps that I generated using cone-isolating stimuli, I was also able to compare the relative weights of the cone inputs. This analysis showed that cortical red-green cells tended to have balanced cone inputs: they were excited by one class of cones to the same extent that they were suppressed by an opponent class of cones. Such cells likely represent the cortical machinery that subserves Hering's red-green opponent system — the "red-green" chromatic axis. This chromatic axis likely works with a blue-yellow (S vs. L+M) axis and an achromatic axis (each represented by different populations of cells) to represent all hues.

In a second study of these cells (Chapter 3), I considered the temporal aspect of the cells' responses. All the color cells studied were strongly suppressed by one of two stimuli used (either L-cone-isolating or M-cone-isolating). In addition, the responses were often biphasic: a stimulus might have caused an initial decrease in firing rate (suppression), but this was often followed by a rebound increase in firing rate. I wondered whether this temporal aspect of the response might subserve temporal, or successive, chromatic contrast. If so, then one would suppose that a given cell's response could be modified depending on the color of an immediately preceding stimulus. As it turns out, the response of a given color cell to its preferred stimulus (either red or green) is increased if the stimulus is immediately preceded by an oppositely colored stimulus. This is in contrast to non-color cells whose response is generally decreased. Thus color cells, which often are Double-Opponent, are well suited to subserve both spatial and temporal color contrast, two key features of color vision.

This may sound a bit circular — after all, I am defining these cells as "color cells", so why wouldn't they be fit to subserve key aspects of color vision? But it should be noted that there is some controversy about whether or not cortical cells are actually responsible for encoding single stimulus features, such as color or motion or form. This will be discussed further (Chapter 2§4); suffice to say here that the recordings presented here support the notion that some cells selectively care about some stimulus attributes and not others.

Chapter 2

Spatial structure of cone inputs to cortical color cells

We need not examine the explanations of color constancy by Helmholtz and those who have followed him during the last century because . . . the paradox [of color constancy] does not really exist: The color of an object is not determined by the composition of the light coming from the object.
— Edwin Land, 1977

(then by what is color determined?)

2§1 SUMMARY AND CONCLUSIONS

The spatial structure of primate cortical color-cell receptive fields is controversial. In this study, spots of light that selectively modulate one class of cones (L, M or S, or loosely red, green or blue) were flashed in and around the receptive fields of V-1 color cells to map the spatial structure of the cone inputs. The maps generated using these cone-isolating stimuli, together with an eye-position corrected reverse correlation technique, produced four findings. First, the receptive fields were Double-Opponent, an organization of spatial and chromatic opponency that, when complemented by other cues in the natural world (e.g. spectral highlights and 3-D shape, Kraft and Brainard, 1999), likely contributes to color constancy and color contrast. Optimally stimulating both center and surround subregions simultaneously with adjacent red and green spots excited the cells more than stimulating a single subregion. Second, red-green cells responded in a luminance-invariant way. For example, red-on-center cells were excited equally by a stimulus that increased L-cone activity (appearing bright red) and by a stimulus that decreased M-cone activity (appearing dark red). This implies that the opponency between L and M is balanced and argues that these cells are encoding a single chromatic axis. Third, most color cells responded to stimuli of all orientations and had circularly symmetric receptive fields. Some cells, however, showed a coarse orientation preference. This was reflected in the receptive fields as oriented Double-Opponent subregions. Fourth, red-green cells often responded to S-cone stimuli. Responses to M- and S-cone stimuli were usually aligned, suggesting these cells might be best described as red-cyan. In sum, red-green (or red-cyan) cells, along with blue-yellow and black-white cells, establish three chromatic axes that are sufficient to describe all of color space.

2§2 METHODS

2§2.1 General Design

Experiments were done in alert adult male macaque monkeys. Macaques are a useful model for human color vision because psychophysical results in them match those of humans (De Valois et al., 1974; Sandell et al., 1979). Moreover, the psychophysical results on human color matching are well predicted from the spectral sensitivities of the macaque cones (Baylor et al., 1987). Monkeys were trained to fixate within a one-degree radius of a fixation spot to receive a juice reward. Results collected when the monkeys moved their eyes outside this tight fixation window were not analyzed. Eye position was monitored with a scleral eye coil and field-coil monitoring system (CNC Engineering, Seattle, Washington), with a spatial resolution of 0.05 degrees, which was calibrated at the beginning of each recording session by having the monkeys look at the center of the monitor and 4 dots at the corners of the monitor. The monkeys had to maintain fixation for 3-4 seconds within the fixation window to receive a juice reward. During periods of stable fixation, average residual eye movements were less than 0.25 degrees. These eye movements were compensated for using an eye-position correction technique (Livingstone et al., 1996). In generating one-dimensional space-time maps, this technique makes it possible to measure receptive field widths as narrow as 0.2 degrees (Livingstone and Tsao, 1999), a resolution finer than the receptive field subregions of the cells studied here (color cells typically had centers around 0.5 degrees wide).

Neuron responses were recorded extracellularly using fine electropolished tungsten electrodes coated with vinyl lacquer (Frederick Haer, Bowdoinham, Maine; Hubel, 1957). Units were isolated using a dual-window discriminator (BAK Electronics, Germantown, Maryland) after they were amplified and bandpass filtered (1-10 kHz). I analyzed only well-isolated units. This recording technique may select for less densely packed neurons, or neurons with larger somas, though it is (at present) impossible to say if this is the case. My implicit assumption is that I am able to sample all neurons without bias.

Most of the color cells I studied had receptive field centers larger than 0.3 degrees. Cells resembling geniculate Type I cells, which likely reside in layer 4Cβ (Livingstone and Hubel, 1984), have tiny receptive field centers (less than 0.2 degrees at the eccentricities recorded here). I generally avoided mapping such cells for two reasons. First, it was hard to maintain a stable recording of such cells for the extended periods required to generate satisfactory spatial maps [perhaps because these are small cells packed densely in layer 4Cβ (Livingstone and Hubel, 1984)]. And second, there is some debate about the validity of cone-isolation with the small spots required to map their small receptive fields. Small spots, more than large spots, will be blurred by chromatic aberration, which will undermine the cone-isolation.

All cells were tested with manually guided spots and bars prior to quantitative receptive field mapping. From this preliminary mapping I excluded cells with very small receptive field centers, cells that might be cortical Type I cells. However, I did collect quantitative maps for a few such cells, in the interests of comparing them to the cells with Type II-like centers. These spatial maps make it clear that cortical Type I cells represent a distinct class of color cells (see Figure 2.11).

To identify and map color cells, I tested cells with stimuli that selectively modulate a single class of cones. Cone-isolating stimuli can be made with the silent substitution method of Rushton (Donner and Rushton, 1959; for review see Estevez and Spekreijse, 1982). Typically such stimuli involve a constant gray adapting background on which stimuli either increase (a plus stimulus) or decrease (a minus stimulus) the activity of a given cone class. The gray background provides an adapting field that has the advantage of being the same for all cone-isolating stimuli but the disadvantage that it limits the cone contrast that can be achieved. To boost the cone contrast in the present study, I used cone-isolating stimuli that had different adapting backgrounds depending on the cone to be isolated. The validity of using such stimuli was explicitly demonstrated by the fact that the results are comparable to results obtained using cone-isolating stimuli presented on a constant gray background (see Figure 2.8). A plus stimulus for a given cone class consisted of a small patch of light that selectively activated that cone (+ state) surrounded by a field of light that selectively inactivated that cone (- state) (see Table 1). A further advantage of using the high-cone contrast stimuli is that they employ a higher mean luminance than conventional stimuli presented on gray backgrounds, which helps saturate rods [e.g., (0, 252, 0), which is the background to the M-minus stimulus, has a luminance of ~60 cd/m2; compared to (175, 143, 126), the background gray, has which a luminance of ~27 cd/m2).

2§2.2 Generating cone-isolating stimuli

I used six stimuli: L-plus, L-minus, M-plus, M-minus, S-plus and S-minus. Each stimulus consisted of a small patch (~0.5° diameter) of one color that was surrounded by a full field (~16° width) of a different color. Between these two colors, only the activity of the desired cone was modulated: the plus stimuli increased the given cone's activity, and the minus stimuli decreased it.

Each color was defined by an RGB state. The L-plus stimulus, for example, consisted of a small patch of (L+) state (255, 0, 0) surrounded by a field of (L-) state (0, 154, 38) (Figure 2.2). The L-minus stimulus consisted of a small patch of (L-) state (0, 154, 38) surrounded by a field of (L+) state (255, 0, 0) (Table 1). These states were generated as follows. The emission spectra for the 3 guns (RGB) of the computer monitor used were determined separately using a Photo Research PR 650 SpectraScan spectrophotometer (Chatsworth, Calif.). I then calculated a 3 x 3 matrix (shown below) representing each cone type's activity due to each gun at 255 by taking the dot product of these spectra with the cone fundamentals (Smith and Pokorny, 1972; Smith and Pokorny, 1975), sampled every 4nm.

$$
\begin{pmatrix} L_{cone} \\ M_{cone} \\ S_{cone} \end{pmatrix} = \begin{pmatrix} 129.7517 & 131.6641 & 50.6336 \\ 26.4562 & 83.8079 & 46.1203 \\ 2.4618 & 6.0469 & 178.0737 \end{pmatrix} \begin{pmatrix} R_{phos} \\ G_{phos} \\ B_{phos} \end{pmatrix}
$$

Using this matrix I calculated relative gun values that would yield the two states [(+) and (-)] for each cone class (Table 1).

Table 1. Phosphor values and cone activites of the high cone-contrast cone-isolating stimuli

	(+) state			(-) state		
	L/R	M/G	S/B	L/R	M/G	S/B
L-isolating						
relative gun	100	0	0	0	31.4	0.32
relative cone	12975	2646	246	4150	2646	246
gun (cd/m2)	44.32	0	0	0	18.9	0.08
gun (0-255)	255	0	0	0	156.8	55
cone activation	130.956	27.055	2.652	43.484	27.128	2.679
final gun	255	0	0	0	154	38
M-isolating						
relative gun	0	100	0	100.7	0	2.005
relative cone	13166	8381	605	13166	2756	605
gun (cd/m2)	0	59.63	0	44.32	0	0.51
gun (0-255)	0	254.3	0	255	0	62.6
cone activation	131.143	83.479	6.043	131.312	27.500	6.004
final gun	0	252	0	255	0	55
S-isolating						
relative gun	64	0	255	0.9011	160.2	0
relative cone	21216	13454	45566	21216	13454	0.0971
gun (cd/m2)	11.12	0	25.6	0.157	37.735	0
gun (0-255)	143.6	0	255	56.3	211.1	0
cone activation	86.904	54.218	181.626	87.142	54.629	3.933
final gun	148	0	255	52	209	0

Each cone-isolating stimulus (L-isolating, M-isolating, and S-isolating) consisted of two states, (+) and (-), such that between the two states only the desired class of cones was modulated. Relative phosphor (R,G,B) and cone activities (L,M,S) were derived (see Text), and these were then normalized [gun (cd/2)] and converted to standard 0-255 values using the gun luminance functions (Figure 2.1).

The method for developing high cone-contrast cone-isolating stimuli was simple. For the L-stimulus, I used the maximal red phosphor during the (L+) state and then matched the activity of the M and S cones produced by the red phosphor with the green and blue phosphors during the (L-) state. I ended up with 6 relative gun values: 3 for the (L+) state and 3 for the (L-)

state. A similar method was used to generate the gun values for the M and S (+) and (-) states, summarized as "relative gun" in Table 1.

These gun values cannot be used directly because they assume that the luminance function for each gun is linear, which is not the case (Figure 2.1). The relative gun values were converted to luminance values (gun (cd/m²) in Table 1) and then to the conventional 0-255 values (gun (0-255) in Table 1) using polynomials (lines in Figure 2.1) fit to the empirically derived gun luminance functions (circles, crosses and triangles, Figure 2.1). To verify that these derived values were actually cone isolating, I measured the emission spectra (separately) for each state.

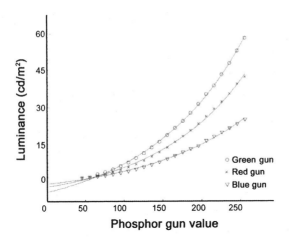

Figure 2.1. Gun luminance functions of the monitor used.

I calculated the cone activities elicited by these two states by taking the dot product of their emission spectra with the three cone fundamentals. This is analogous to the approach used by others (Chichilnisky and Baylor, 1999). A given stimulus always modulated the desired cone class much more than it modulated either of the other two cone classes. To improve them further, though, I adjusted the gun values of the two states, and recalculated the cone activities until the undesired cone classes had modulation indices of less than 0.4. These adjusted gun values and the resulting cone values are listed as "final gun" and "cone activation" (Table 1). The cone modulation index =

[(max. cone activity − min. cone activity)/(max. cone activity + min. cone activity)] * 100.

The L stimuli had a modulation index of 50.1, M of 50.4, and S of 95.8. The cone isolation was checked periodically (every month or so) and appropriate minor changes made to the phosphor values. A final confirmation that the L-cone stimulus was cone-isolating was obtained by presenting it to a protanope who found it almost invisible.

Under the stimulus conditions used, a significant contribution by the rods is unlikely because the rods are probably saturated. The luminance (in cd/m^2) for each state was roughly L(+), 44; L(-), 19; M(+), 60; M(-), 45; S(+), 37; and S(-), 38. The illuminance limit for human rods is 1800 scotopic trolands (Hess and Norby, 1986) which is equivalent to 7200 photons per receptor per second (Spillman and Werner, 1990). The luminance of the S-cone stimulus (37 cd/m^2) corresponds to ~ 10400 photons absorbed per receptor per second (cd/m^2 * 10 * pi * radius of pupil $(mm)^2$ (pupil is ~3mm)), putting the S-cone stimulus securely in the photopic range. Qualitatively, all the stimuli appear bright and vividly colored.

I also generated conventional cone-isolating stimuli presented on a constant adapting background (e.g. Reid and Shapley, 1992). Cone modulation indices for these stimuli were L = 33.3, M = 41.2 and S = 94.7. The constant gray background was (R, G, B) 175, 143, 126; ~ 27 cd/m². The results using these stimuli were comparable to those using the high cone-contrast stimuli (see Figure 2.8 and Figure 2.14).

2§2.3 Stimulus presentation for receptive-field mapping

Stimuli were presented (in a dark room) on a computer monitor (Barco Display systems, Kortrijk) 100 cm from the monkeys' eyes. The stimulus used to generate a given map involved presenting a single small patch of cone-isolating light (~ 0.4 x 0.4° square) at random locations in and around the cell's receptive field while the monkey fixated (Figure 2.2). The adapting background covered the full 21" monitor (about 20° of visual angle). Because the adapting backgrounds were different for each stimulus in a stimulus pair (e.g. L-plus and L-minus have different adapting backgrounds, see Table 1), the two maps for each pair were generated separately from different stimulus runs (with cone-isolating stimuli presented on a constant gray background the plus and minus responses are usually generated from a single stimulus run). The size of the stimulus patch was optimized for each cell. Stimuli were presented for 30-100 ms in each location, and there was a 13 ms refresh delay before presentation in a new location. The size of the stimulus was not critical, as long as the stimulus was smaller than the receptive field center. I confirmed this by comparing maps obtained from the red-on-center cell shown in Figure 2.6A (using 0.4 x 0.4° square stimuli) with maps obtained using smaller stimuli (0.15 x 0.15° square). Similar receptive field sizes were obtained.

Maps reflect stimulus positions 50-70 ms before each action potential and are smoothed with a gaussian filter. The 50-70 ms delay corresponds to the visual latency of the cell. Each map is an average of at least 40 (and usually many more) presentations everywhere in the receptive field. Each map took

36

from 15-45 minutes and usually consisted of at least 1500 spikes.

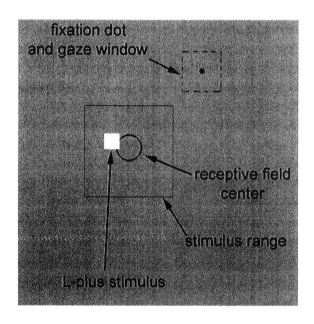

Figure 2.2. The L-plus stimulus used for mapping the 2-D spatial receptive field. In this grayscale reproduction, the light gray L-plus square represents the red square that appeared on the computer monitor and the darker gray background, the darker green background. The boxes demarking the stimulus range and the fixation window were not visible to the monkey.

2§2.4 Note concerning the S cone-isolating stimulus

The cone-isolating stimuli were based on the cone fundamentals of Smith and Pokorny (1972, 1975). These fundamentals are derived from color matching functions for the central 2 degrees. The corresponding region of the retina is protected by macular pigment. Macular pigment, which absorbs a significant amount of light of shorter wavelengths, is densest in the fovea, and falls off with eccentricity (Polyak, 1957). Cone-isolating stimuli based on Smith and Pokorny (1972, 1975) are therefore best applied to cells whose receptive fields are in the central 2 degrees. Macular pigment interferes minimally with L and M cone isolation because the macular pigment absorbs in the shorter wavelengths (400-500 nm) and L and M cone-isolating stimuli employ the red and green phosphors which emit scarcely in the shorter wavelengths. Macular pigment may, however, pose a problem in interpreting responses to S cone-isolating stimuli if the receptive fields are outside the macular-pigment region of the retina. To test this, I mapped two cells having receptive fields at 5 degrees, outside the macular rich region, with cone-isolating stimuli that used cone fundamentals derived from color matching at 10° eccentric to the fovea (Stockman and Sharpe, 2000). The maps were qualitatively identical to those generated using cone-isolating stimuli based on Smith and Pokorny (1972, 1975). All cells recorded were between 2.5 and 5 degrees eccentricity.

One might have thought that lateral chromatic aberration (the displacement on the retina of the blue image resulting from the greater refraction of short wavelength light) would interfere with cone isolation, but this probably is not a problem, for two reasons. First, assuming the monkeys have pupils that are approximately aligned with the visual axis, as is so for the average human, lateral chromatic aberration would account for shifts of less than 1.3 arc minutes at 5 degrees eccentricity (Thibos et al., 1990). (All cells studied were within 5 degrees of the fovea.) This amount of shift is small compared to the diameter of the receptive field centers of the color cells (~ 0.5 degrees). Second, there was no systematic shift of the blue map for cells in a given receptive-field location, as would be expected if lateral chromatic aberration were underlying the spatial shifts between the maps.

The effects of longitudinal chromatic aberration should also be considered. Given that the eyes are generally focused in the yellow, the S-cone stimulus — which employs predominantly blue light — might be defocused. This longitudinal (or axial) chromatic aberration would have two effects. First it would result in a stimulus spot that was slightly larger than an L or M-isolating stimulus spot of nominally the same size, which would blur the receptive field map of the S-cone. This is not a major problem because the spatial structure of the response to the S-cone stimulus is interpretable: the suppression in the center of a red-on-center/cyan-off-center cell is clear (e.g. Figure 2.6A). Blurring may nevertheless pose some problem for the S-cone stimulus because it would result in a smaller contribution of the shorter wavelengths to the S stimulus than I thought. If blurring caused a significant decrease, I would have overestimated the impact of the (S+) state (which uses maximal blue gun) on the L and M cones, and the resulting activity of the L and M cones would be slightly higher during the (S-) state than during the (S+) state because the G and R phosphors are used during the (S-) state and these phosphors are not affected by blurring. Thus the (S-) state would not only decrease the activity of the S cones relative to their activity during the (S+) state, but would also increase the activity of the M and L cones. This would seem not to be a problem in the present study because I only studied cells that gave opponent responses to M and L-plus stimuli: the increased activity of the L-cones would cancel the effect of the increased activity of the M cones. But the blue gun stimulates the L cones slightly more than it stimulates the M cones (the blue gun activates the L cones by 50.6336 and the M cones by 46.1203 relative units, see matrix above), so that the effect of chromatic aberration during the (S+) state would be slightly greater for the L-cones than the M-cones, and consequently the calculated activity of the L cones would be overestimated slightly more than the calculated activity of the M cones. If this were significant, it would mean that the (S-) state would have acted slightly more like an L-plus stimulus than like an M-plus stimulus. One would expect accordingly red-green cells that do not receive any S cone input to respond (probably weakly) to an S-minus stimulus in a way predicted by the L-plus stimulus. A recent preliminary study contends that the effects of longitudinal chromatic aberration are insignificant (Cottaris et. al, 2000).

2§2.5 Data Quantification

The response maps were quantified by generating post stimulus time histograms (PSTHs) corresponding to stimuli presentation to the most active region of the center and the surround. The sizes of the regions from which spikes were collected were defined by the size of the stimulus used to map each cell; only one site was sampled for the surround response and one site for the center response and the activity was normalized for the number of stimulus presentations. This method of quantification likely underestimates the impact of the surround on the cells' responses because it does not account for the larger spatial extent of the surround than the center. Cell response was determined by subtracting the background rate from the peak. Background measurements were determined as the first 40 ms of the PSTHs. Standard errors of the measurement (peak activity-background activity) are given in Figure 2.12 when they are larger than the size of the symbols.

2§3 RESULTS

2§3.1 Screening for color cells

Cortical cells were screened with cone-isolating stimuli: every cortical cell encountered was tested with alternating flashes of a large spot (~ 1 x 1°) of L-plus and M-plus cone-isolating light centered on the receptive field. The patch was manually guided in and out of the receptive field to find the cell center. (Cells were also tested with spots and flashing and moving oriented bars of various colors.) A cell was designated a color cell if it responded vigorously to small spots of colored light and if the L-plus and the M-plus stimuli produced opposite responses (excitation vs. suppression; e.g. Fgure 2.3). I screened for red-green cells because, in the cortex, they are ~8 times more common than blue-yellow cells (Wiesel and Hubel, 1966; Livingstone and Hubel, 1984; Ts'o and Gilbert, 1988). Simple luminance cells (cells lacking cone opponency) would produce responses similar to those of the M-plus and L-plus stimuli. Complex cells, which are by far the major cell type in V-1, produced weak and transient responses at the onset and offset of all cone-isolating stimuli. I screened ~615 single units in the primary visual cortex to obtain 73 red-green color cells (8 of the 73 were discarded because they were Type I cells, see Figure 2.11). All cells were between 2.5 and 5 degrees eccentricity. Most showed a complete suppression of firing to one of the two stimuli (e.g. Figure 2.3) making them easily recognizable on an audio monitor. Cells were determined to be either L-plus on (i.e. red-on center), or M-plus on (i.e. green-on center). 36/65 were red on, 29/65 were green on.

I should note in passing that since the completion of this study I have screened for and mapped several blue-yellow cells. The maps of these cells will not be presented; suffice to say, that these few blue-yellow cells gave responses to L stimuli that were similar (either excitation or suppression) to those that they gave to M stimuli, and strong opponent responses to the S-cone stimulus. Curiously, the L and M inputs to the blue-yellow cells were not balanced; one cell, for example, gave a very weak response to the M stimulus and a strong response to the L stimulus and another gave a strong response to the L stimulus and a weak response to the M stimulus. One might be surprised at this result, but the L and M inputs need not be balanced for these cells to subserve the blue-yellow axis, because the yellowish-bluish cardinal direction corresponds to the tritanopic confusion line and not to the

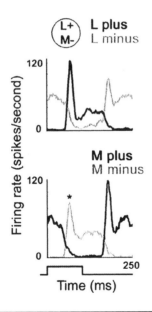

Figure 2.3. Responses of a cortical color cell, having an L+/M-center, to large spots of cone-isolating light that encompassed the receptive field center. Stimulus duration is indicated by the step at bottom.

classical red-green equilibrium line (Krauskopf et al, 1982), and these cells clearly could encode the tritanopic confusion.

Using physiological criteria, color cells were found both above and below layer 4C. Most of the cells were strongly monocular (Figure 2.4), consistent with previous reports (Michael, 1978). Post stimulus time histograms (PSTHs) generated using screening stimuli were collected for 47 cells (Figure 2.5); 22/47 cells were red-on center (open squares in Figure 2.5); 25/47, green-on center (filled circles in Figure 2.5). Because the recordings were not intracellular, I cannot distinguish between the specific mechanism of opening chloride channels (i.e. inhibition) and withdrawl of excitation. For this reason, I prefer the less specific term "suppression" (rather than "inhibition") to describe a decrease of a cell's activity in response to a stimulus because this term encompasses both possible causes.

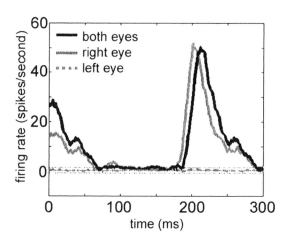

Figure 2.4. Color cells tend to be monocular. This red-on-center cell was stimulated with an L-plus screening stimulus to both eyes, and either eye alone.

2§3.2 Spatial maps of the cone inputs
Testing the surround for double opponency

After identifying a color cell using center-sized spots (Figure 2.3), I mapped it with patches of cone-isolating stimuli that were smaller than those used for screening for color cells. The stimulus for mapping the L-plus response, for example, looked like a single small patch (smaller than the receptive field center) of bright red light flickering on a full field of darker bluish-green in and around the receptive field (Figure 2.2).

Complete mapping included six maps: L-plus, L-minus, M-plus, M-minus, S-plus and S-minus. If time permitted, I also collected the two luminance maps: the map of black on a white background and the map of white on a black background. In addition, to demonstrate the validity of using stimuli presented on differing adapting backgrounds, I also tested several cells with stimuli presented on gray backgrounds ("low cone-contrast stimuli", see Figure 2.8 below). Two-dimensional spatial maps of the receptive fields were generated using the eye-position-corrected reverse-correlation technique of Livingstone et al (1996). The response maps reflect the average position of the stimulus before each spike, accounting for the latency of the cell, and reveal the spatial structure of the receptive fields. During mapping, the cell's waveform was monitored to ensure all maps were derived from a single cell. Of the 65 cells screened for color, the complete ensemble of maps was collected for 24 cells. In 25 of the remaining cells at least 3 of the 6 maps were obtained, permitting an assessment of the cone inputs to the surround; the data from these cells supports the conclusions based on the 24 cells with complete maps. Three red-on-center cells with complete maps are shown in Figure 2.6A,B,C; one green-on-center cell with complete maps is shown in Figure 2.6D.

44

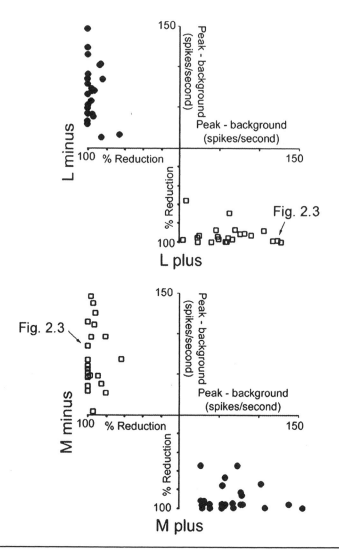

Figure 2.5. Responses determined from PSTHs for 47 color cells screened as in Figure 2.3. Red-on-center cells (open squares); green-on-center cells (filled circles). The response was categorized as suppression (e.g. the response to L-minus in Figure 2.3) or excitation (e.g. the response to L-plus in Figure 2.3). (*continued*)

(*Figure 2.5 continued*) Suppression was then quantified as a percentage of reduction of background, in which background was calculated from the firing rates for the first 40 ms of the PSTH. Excitation was quantified as (peak - background) in spikes/second. The cell whose PSTH is given in Figure 2.3 is identified for reference.

As expected for a red-on-center cell, the receptive field center was excited by the L-plus stimulus (upper left panel, Figure 2.6A). The center was chromatically opponent, as shown by the suppression by M-plus spots in this region (middle left panel, Figure 2.6A). The S-plus stimulus presented in the center of the receptive field was also suppressive (bottom left panel, Figure 2.6A). Suppression in the plus stimulus maps can be inferred by the minus stimulus maps, which reveal a similar spatial distribution but produce excitation. For example, the center, which was suppressed by M-plus, was excited by M-minus. Two other red-on-center cells at a similar eccentricity are shown (Figure 2.6B,C). All three cells were Double-Opponent: L-plus, M-minus and S-minus excited these cells in the center of their receptive fields; in spatial and chromatic opponency, these cells were excited by L-minus, M-plus and S-plus in the surround. This pattern of response is summarized in Figure 1.6D.

One might have thought that the surround response to M-plus, S-plus and L-minus is simply due to the higher background firing-rates that these conditions elicit (because the color of the background of these stimuli would drive the center of the receptive field); but this cannot be the case because the surround responses are not uniform, which they would be if the surround responses were non-specific. Thus the surround response in these cells must be attributed to stimulating the surround of the cells' receptive fields.

Surround strength varied from cell to cell. A red-on-center cell with a strong surround (Figure 2.6C) and a green-on-center cell with a very weak surround illustrate this (Figure

2.6D). Note that the response maps represent the actual firing-rate and are <u>not</u> corrected for the variability in background firing-rate between stimulus conditions. Thus the overall brightness of each map varied. This can be misleading because the elevated background activity produced by some of the stimulus conditions can be misinterpreted as a strong surround (e.g. Figure 2.6D). To be significant, the brightness of the map must drop off in the periphery, as is clear for the cell in Figure 2.6C, but not as clear for the cell in Figure 2.6D.

The variability between backgrounds was accounted for in quantifying the responses (see Figure 2.12 below). In earlier studies, the cell shown in Figure 2.6D might have been called a Type II cell (a cell having no surround), or a ¾ Double-Opponent cell (a cell having a surround fed by a single cone class). But the mapping technique used here allowed a quantitative assessment of the cone inputs to the surround, and though they are weak, they are chromatically opponent (Figure 2.7). The excitation by L-plus in the surround is clear as a peak in the post stimulus time histogram at about 70 ms (Figure 2.7 top PSTH, gray trace). The suppression by M-plus in the surround is evident as a dip (Figure 2.7 middle PSTH, gray trace).

48

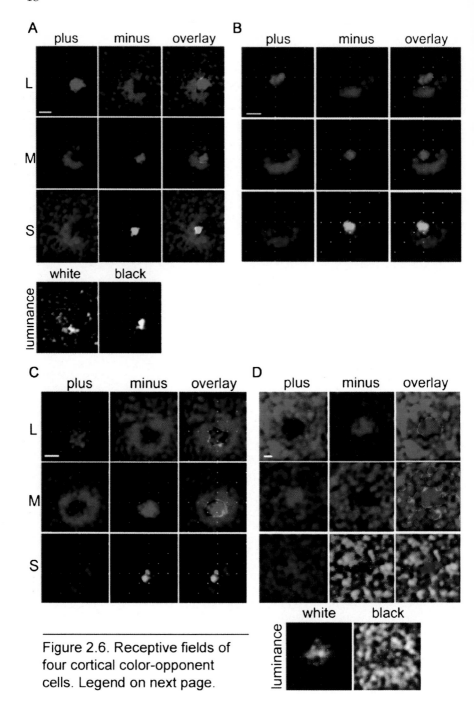

Figure 2.6. Receptive fields of four cortical color-opponent cells. Legend on next page.

←
Figure 2.6. Receptive fields of four Double-Opponent cells recorded in alert macaque V-1.

A. A small patch of cone-isolating light was flashed at random locations in and around the receptive field; response maps were generated using an eye-position reverse correlation technique (see Chapter 2§2). The maps reflect the average stimulus position that preceded each spike, and are corrected for the visual latency. L-plus, M-plus and S-plus (left column); L-minus, M-minus and S-minus (middle column); and overlay (right column). The background firing-rate has not been subtracted from these maps. Black in these maps represents a firing-rate of zero spikes/second; more intense responses are represented as more saturated colors. Peak firing-rates (spikes/second) were L-plus = 62, L-minus = 20, M-plus = 35, M-minus = 47, S-plus = 16, S-minus = 71, white = 21, black = 27. Scale bars are 0.5°; stimulus size was 0.4 x 0.4°. This cell was 4° peripheral.
     The coloring of the maps does not match that of the stimuli. The L-plus and M-minus maps are colored red because the stimulus in both cases appears red: the L-plus stimulus is a bright red (on a dark bluish green), while the M-minus is a dark bluish-red (on a bright green background), see Table 1.

B. Response maps for a second red-on center Double-Opponent cell. Peak firing-rates (spikes/second) of L-plus = 67, L-minus = 45, M-plus = 77, M-minus = 57, S-plus = 23, S-minus = 30. This cell was 5° peripheral. Stimulus size was 0.4 x 0.4°.

C. Response maps for a third red-on center Double-Opponent cell. Peak firing-rates (spikes/second) were L-plus = 53, L-minus = 37, M-plus = 55, M-minus = 71, S-plus = 10, S-minus = 54. This cell was 5° peripheral. Stimulus size was 0.4 x 0.4°.

D. Response maps for a weakly Double-Opponent green-on center cell. Post-stimulus time histograms corresponding to stimuli presentations in center and surround are shown for this cell in Figure 2.7. Peak firing-rates (spikes/second) of L-plus = 23, L-minus = 43, M-plus = 41, M-minus = 12, S-plus = 53, S-minus = 37, white = 32, black = 22. This cells was 5° peripheral; stimuli were 0.4 x 0.4°.

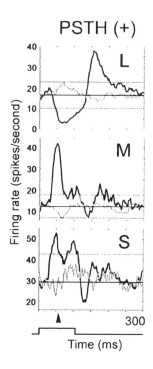

Figure 2.7. For the cell shown in Figure 2.6D, responses were determined from the reverse correlation data by selecting segments of the spike train corresponding to presentations of plus stimuli in the center (black traces) and surround (gray traces). Stimulus duration was 100 ms (indicated at bottom). One standard deviation above and below the mean background firing-rate is given for reference. The response maps (Figure 2.6D) show the spatial structure of the response between 50 and 70 ms after the onset of the stimulus (arrowhead).

To quantify the maps (see Figure 2.12 below), the "surround responses" (e.g. the PSTHs shown in Figure 2.7) were obtained from the most sensitive region within a ring one receptive-field-center-width wide, surrounding the center, and were only deemed significant if they were greater than responses to stimulation of regions well outside the receptive-field surround. Some cells showed very weak surrounds (see Quantification of spatial structure, Figure 2.12), but all cells (with significant surround responses) showed opposite responses within each receptive field subregion for every stimulus pair (plus and minus).

Time permitting, the luminance maps were collected (e.g. gray maps of Figures 2.6A and 2.6D). A perfect balance of opponent cones would predict no response to broadband light. Some Double-Opponent cells, however, were not perfectly balanced. This was the case for the red-on-center cell (Figure 2.6A) which gave a response to black. Though some color cells have been reported to respond to luminance (Johnson et al, 2001), it is difficult to interpret my maps without making assumptions about how the cone inputs are summed, especially given that the "white" stimulus (all guns set to 255) did not appear perfectly white and did not modulate the cones in a way predicted by the sum of the three separate plus stimuli. Instead, the cone interactions were tested directly with a different set of experiments (see Figure 2.14, below).

It has been suggested that cone-isolating stimuli presented on gray backgrounds are not effective for driving cortical color cells because of the limitation on cone contrast (David Hubel and Margaret Livingstone, personal communication; see Methods). To overcome this potential problem, high cone-contrast stimuli were used for spatial mapping (e.g. Figures 2.6 and 2.7). I also mapped a few cells with stimuli presented on constant adapting gray backgrounds ("low cone-contrast stimuli") to test the validity of using the high cone-contrast stimuli. Such low cone-contrast stimuli elicited responses, but the maps from the high cone-contrast stimuli were typically clearer (Figure 2.8A). For example, the annulus of excitation produced by the M-plus stimulus for a red-on-center cell is readily evident for the high cone-contrast map, but not so clear for the low cone-contrast map (Figure 2.8A). For this cell, the spatial opponency was only revealed with the low cone-contrast stimuli when the minus response map was subtracted from the plus response map (Figure 2.8B), making the low cone-contrast stimuli less suitable for a direct assessment of the spatial receptive field structure. In other cells the spatial structure is just as clear in both low and high-cone contrast maps.

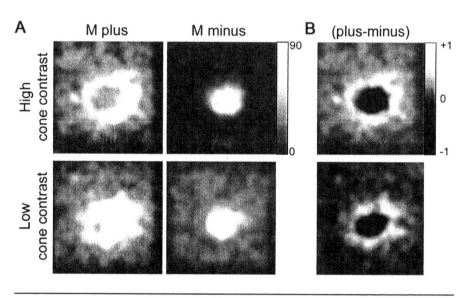

Figure 2.8. Comparison of response maps generated using high cone-contrast stimuli and low cone-contrast stimuli.
A. Response maps for the red-on-center cell shown in Figure 2.6C were generated using both high cone-contrast stimuli (upper panels) and low cone-contrast stimuli (lower panels). Only the maps for M-plus and M-minus are shown. The responses are coded with a linear grayscale firing-rate bar, black represents 0 spikes/second and white represents 90 spikes/second. All maps reflect responses after the same number of stimulus presentations. Scale bar = 0.5°.
B. Difference maps between the minus and plus response maps. Middle gray represents no difference between the minus and plus maps. (*continued*)

(*Figure 2.8B continued*) Note that the region surrounding the receptive field for the low cone-contrast stimulus is roughly middle gray reflecting the constant background activity between the plus and minus conditions. This is not the case for the high cone-contrast stimuli, which have different colored backgrounds and therefore different background firing-rates for the plus and minus conditions.

In addition to using the low-cone contrast stimuli to map the 2-dimensional spatial structure of the cone inputs of a few cells, the low cone-contrast cone-isolating stimuli were also used to map the responses to simultaneous and sequential color spots (e.g. the responses to an L-cone spot next to an M-cone spot, and the responses to an L-cone spot following an M-cone spot) of 36 cells. These experiments are described in Chapter 2§3.6 and Chapter 3§3.4. It should be noted here, though, that 16/36 of these color cells in these experiments, which employed low cone-contrast stimuli, did not give obvious surround responses (e.g. cells shown in Figure 3.3E, F), and are perhaps best described as cortical Type II cells (see Figure 1.6). It is possible that with higher cone-contrast stimuli these cells might have given stronger surround responses.

Another green-on-center cell is shown in Figure 2.9. The double opponency is clear as an increase in the L-plus, M-minus and S-plus maps as a ring of increased activity surrounding a pit of suppression. In contrast to the cells presented in Figures 2.6, the S-cone input in this cell aligns in space and sign with the L-cone input. The S-plus stimulus produced remarkably potent suppression in this cell, lasting almost twice the stimulus duration (Figure 2.10).

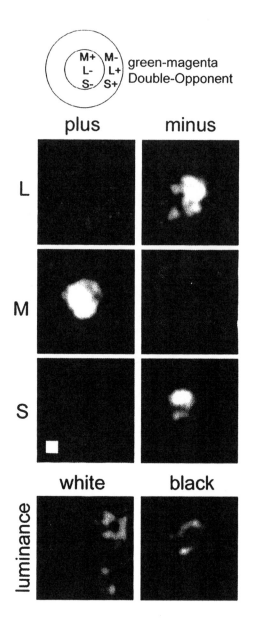

Figure 2.9. A green-magenta cell. Higher firing-rates are given by whiter regions; black represents zero spikes/second. Peak firing-rates for this cell are given Figure 2.10. Stimulus size was 0.4x0.4° (white box).

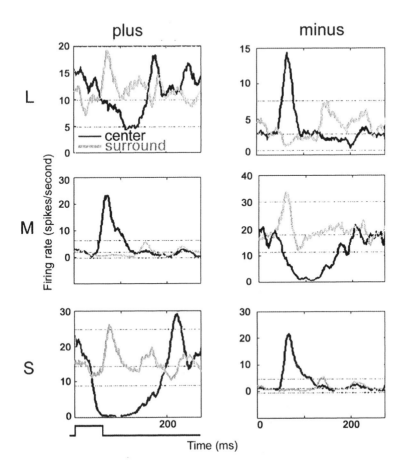

Figure 2.10. Post stimulus time histograms (PSTHs) corresponding to stimulus presentations to the center and surround of the green-magenta Double-Opponent cell shown in Figure 2.9.

The spatial maps had sufficient resolution to assess whether the cone contribution to the surround was homogeneous. Often the surrounds were non-uniform; for example the contribution of the M cones to the surround of the red-on-center cells shown in Figure 2.6A,B were not perfectly annular but crescent-shaped. The full range of surrounds exhibited by Double-Opponent cells is shown by comparing the extent of the surround in Figure 2.6C (a full annulus), in Figure 2.6A (a crescent) and below in Figure 2.14A (an adjacent, oriented field). Nine out of 49 cells showed a receptive-field organization similar to that of the cell shown below in Figure 2.14A, where the subregions were not suitably described as center and surround but rather as adjacent subregions. Moreover, these adjacent subregions were themselves coarsely oriented, as is the case for the S-cone subregion, Figure 2.14A.

2§3.3 Distinguishing color cell types:
Spatial receptive fields of Type I cells

Type I and Type II cells, found in the lateral geniculate nucleus, and Double-Opponent cells, found in the cortex, all give opponent responses to different classes of cones but the spatial structure of the cone inputs and the size of the receptive fields are different (Figure 1.6; Livingstone and Hubel, 1984). A Type I cell receptive field has a small center, which for cells representing the fovea, might be fed by a single cone; the opponent input encompasses a larger spatial extent, including the center, such that the receptive field surround is fed by a single cone class (Figure 1.6A). A Type II cell receptive field consists of a single relatively large region in which the cone inputs are opponent and co-extensive; a Double-Opponent cell receptive field is analogous to that of a Type II cell with the addition of a chromatically opponent surround (Figure 1.6). All three cells types can be distinguished from non-color cells by their responses to the relatively large patches of screening stimuli (Figure 2.3 and 2.5) and, as I will show in Chapter 3, by their spatial cone-interaction maps (see Figure 2.14).

Using the relatively large sized stimuli spots used to screen for color cells, Type I cells would be indistinguishable from Type II cells. To make the distinction between Type I and Type II cells (if in fact Type I cells exist in the cortex) one must consider the 2-D spatial receptive field, and use much smaller stimulus spots. In general this was assessed qualitatively using manually guided spots and bars. I did not generally collect quantitative spatial response maps for cells that I thought were Type I cells (for the reasons discussed in 2§2.1). But in a few cases I did intentionally map what I thought were Type I cells (Figure 2.11), in the interests of comparing them to cells with larger receptive field centers.

It is worth noting that, in the cortex, no one has mapped the spatial structure of the cone inputs to Type I cells, though such cells probably exist (Livingstone and Hubel, 1984). Thus my designation of this class of cells as Type I is based solely on a comparison of my spatial maps of them with the spatial maps of Type I cells in the lateral geniculate nucleus (Reid and Shapley, 1992). To facilitate this comparison, I generated the spatial maps using stimuli presented on gray backgrounds because Reid and Shapley used gray backgrounds. Note, however, that stimuli presented on gray backgrounds and stimuli presented on different adapting backgrounds give similar results (Figure 2.8).

The spatial maps of the Type I cells (Figure 2.11A), as with the spatial maps presented above, represent the average response to stimuli presented in and around the receptive field. The responses are coded where black represents zero spikes per second; white, 50 spikes/second; and the middle gray represents the average background activity (~30 spikes/second) (Figure 2.11A). The L-cone stimulus consisted of two spots presented simultaneously on a gray background, a brighter red spot (L-plus) and a darker bluish-green spot (L-minus). The M-cone stimulus consisted of a brighter green spot (M-plus) and a darker bluish-red spot (M-minus).

58

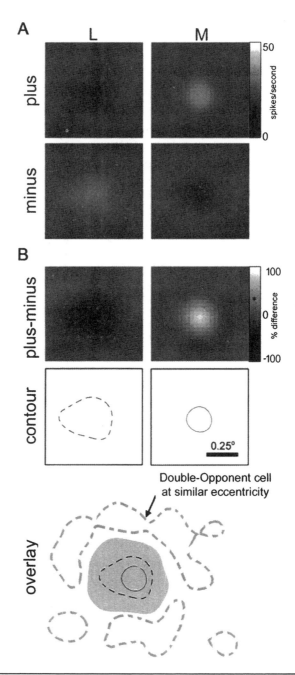

Figure 2.11. Receptive field map for a cortical Type I cell. (*con't*)

(Figure 2.11, continued). Spatial receptive field map for a cortical Type I cell.

A. The response is coded according to the grayscale, peak response (most white) is 50 spikes/second. Note that the cell is color opponent, giving opposite responses to L-plus and M-plus stimuli, and opposite responses to L-minus and M-minus stimuli.

B. Difference maps. The response maps to the minus stimuli were subtracted from the response maps to the plus stimuli. Regions outside the receptive field are roughly middle gray ("0" on the scale bar), indicating no difference between the plus and minus maps. This shows that the background activity for all the stimuli is roughly the same (one can also see this in the raw maps, where the backgrounds are all about the same gray, indicating the cell showed the same average activity under all stimulus conditions).

Contour plots were drawn at +2 and -1 standard deviations from zero difference (asterisks) and are useful in comparing the relative extent of the cone inputs. If this cell were recorded from the LGN, it would be called an M-plus-center/L-minus surround Type I cell because the M cone contribution is slightly smaller than the L cone contribution — this is actually clearest in a comparison between the L-minus map and the M-plus map. The +/- 1.5 std. dev. contour plot for a Double-Opponent cell is drawn for comparison.

Scale bar is 0.25°. Both the "Type I" cell and the Double-Opponent cell were ~4° from the fovea.

Thus the plus and minus response maps are derived from the same spike train. This is in contrast to the spatial maps generated using high-cone-contrast stimuli (e.g. Figure 2.6) where the plus and minus maps are derived from different spike trains.

The plus and minus response maps were compared by subtracting the minus map from the plus map (Figure 2.11B).

The background of these difference maps is roughly middle gray reflecting the constant background activity of the cell under all stimulus conditions. Contour plots maps (Figure 2.11B) show that the L cones suppressed the cell over a slightly larger spatial extent than the M cones excited the cell. This is clearest in a comparison between the M-plus map and the L-minus map: the L minus map, representing the spatial extent of suppression by L cones, shows a larger region of activity than that shown by the M-plus map (emphasized in the contour plots). The relative size of the two cone inputs is consistent with maps of lateral geniculate Type I cells (Reid and Shapley, 1992). The contour plot for the receptive field of an L-plus-center/M-plus-surround Double-Opponent cell at a similar eccentricity is shown for comparison (Figure 2.11B). The shaded region represents the spatial extent of the center response of the Double-Opponent cell; thick dashed gray lines, the surround. As with the Type I cell, the Double-Opponent cell receptive field was mapped using low-cone contrast stimuli. The Double-Opponent receptive field is clearly distinguished from that of the Type I cell by its larger size.

2§3.4 Quantification of spatial structure

One way to quantify a cell's response is to evaluate the change in the cell's activity in relation to its background activity (Figure 2.5). The background activity was defined as the activity in response to the adapting background by itself (extracted from the first 40 ms of the PSTHs, Figure 2.3), and varied for each stimulus condition. Most cells were completely suppressed by either the L-plus or the M-plus stimuli (e.g. black trace bottom plot, Figure 2.3). The M-plus stimulus might have been capable of reducing the cell's firing even further (Figure 2.3) but because the cell's activity cannot drop below zero spikes/second one cannot measure the total extent of suppression — at least with extracellular recording. Because most cells showed a reduction to zero firing to one of the two screening stimuli (i.e. the cells rectify, Figure 2.5), it is likely that the stimulus was capable of more suppression than could be measured directly. To find a more meaningful measure of suppression, I assumed that the suppression was equal in magnitude but opposite in sign to the excitation   produced by   the   opposite   contrast   stimulus (Ferster, 1994; Tolhurst and Dean, 1990; this assumption is essentially that color cells sum their inputs in a linear way, which is true, Figure 3.7). For example, the suppression by M-plus would be equal, but opposite in sign, to the peak excitation by M-minus (e.g. asterisk in Figure 2.3).

I plotted the center and surround responses to L modulation and M modulation for all cells having the complete ensemble of maps (see Methods; Figures 2.12A,B). For the center response, all the red-on-center cells (open squares) fall into quadrant 4: the centers were excited by an increase in L-cone activity and suppressed by an increase in M-cone activity. The green-on-center cells (filled circles) fall into quadrant 2: the centers were suppressed by an increase in L-cone activity and excited by an increase in M-cone activity. This distribution indicates that the centers of all cells were chromatically opponent (this is not surprising given the cells were screened for this). The populations swap quadrants when their surround responses are plotted (Figure 2.12B): the red-on-center cells were excited by M-plus and suppressed by L-plus in their surrounds, and the green-on-center cells were excited by L-plus and suppressed by M-plus in their surrounds. That the cells swap quadrants indicates that they were both chromatically and spatially opponent, and justifies the designation Double-Opponent.

62

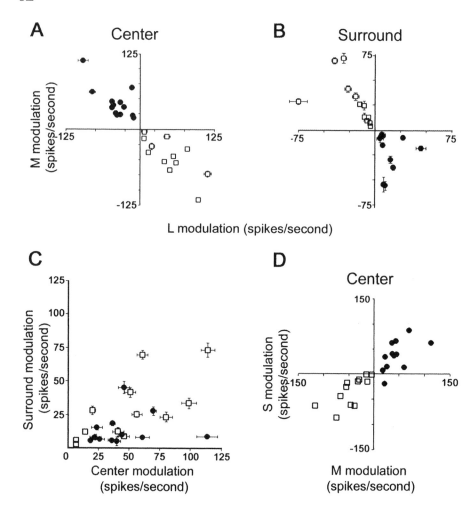

Figure 2.12. Quantification of cone inputs to cortical color cells. Green-on cells (filled circles); red-on cells (open squares).
A and B. Responses were determined from the reverse correlation data by selecting segments of the spike train corresponding to stimulus presentations in the center (A) and surround (B).
C. Surround responses were generally weaker than center responses. (*continued*)

(*Figure 2.12C, continued*) For green-on cells, the L-plus surround response is plotted against the M-plus center response; for red-on cells, the M-plus surround response is plotted against the L-plus center response. The background firing-rate for each measurement has been subtracted from the peak.
D. Red-green cells often responded to the S-cone stimulus.

The strengths of the centers were not equal to the strengths of the surrounds (Figure 2.12C): the surrounds were generally weaker. The contribution of the surrounds to the cells' responses may, however, be underestimated because this measure of surround strength does not take into account the larger spatial extent of the surrounds.

The response to M stimuli when compared to the response to L stimuli has a slope of -0.95, (r^2 = 0.84) (Center, Figure 2.12A) and –0.98, (r^2 =0.53) (Surround, Figure 2.12B). This shows that the cells responded equally well to L-plus stimuli (which look bright red on a darker bluish-green background) and to M-minus stimuli (which look dark bluish-red on a brighter green background). Similarly, the cells responded equally well to M-plus stimuli (which look bright green on a bluish-red background) and to L-minus stimuli (which look dark bluish-green on a bright red background). The cells were thus responding as well to a bright color as they were to a dark color. This luminance-invariant color response was true for both the red-on-center cells and the green-on-center cells (Figure 2.12A). Such a comparison is possible because the L and M cone-isolating stimuli have nearly identical modulation indices (see Methods).

2§3.5 Do Red-Green cells get S-cone input?

In addition to being modulated by the L and the M cone-isolating stimuli, most red-green color cells (45/49, 92%) were modulated by the S cone-isolating stimulus (Figure 2.12D). In most of these (42/45, 93%), S-cone responses were elicited in the same spatial distribution and had the same sign as responses to M cone-isolating stimuli. These cells might tentatively be called red-cyan (cyan = green + blue). The remaining 3 cells with significant responses to the S-cone stimulus had apparent S-cone input aligned with the L-cone input (e.g. Figure 2.9). These might be green-magenta cells.

The similarity in S and M input raises the possibility that the S-cone stimulus was not cone isolating. Of course the fact that M and S responses are correlated (Figure 2.12D) cannot be used by itself to assert that the S cone responses are an artifact: if red-green cells receive S cone input then presumably that input would be well balanced with the other cone inputs in the same way the M and L inputs are well balanced (Figure 2.12A,B). The S-cone stimulus is probably not compromised because of macular pigmentation or lateral chromatic aberration (see Methods). In addition, several lines of evidence suggest that responses to the S-cone stimulus are attributable to the activity of the S cones and not to spillover stimulation of the M cones:

1. The response to the M stimulus did not perfectly predict the response to the S stimulus. Though there was a positive correlation between M and S modulation (slope = 0.75), the r^2 value is only 0.76 (Figure 2.12D). If the S stimulus was simply driving the red-green cells by modulating the M cones (and not because the red-green cells received any S-cone input), then one would expect a higher r^2 value.

2. The S stimulus elicited a stronger response than the M stimulus in one third (8/24) of the cells studied. It is unlikely that a stimulus designed to modulate the S cones would actually modulate the M cones more than a stimulus designed to modulate the M cones.

3. The red-on-center cells frequently responded to the S-plus and S-minus stimuli in ways markedly different from the ways in which they responded to the M-plus and M-minus stimuli. This would not be expected if the S-cone stimulus was simply driving the cells through the M cones. While the centers of red-on-center cells were suppressed and their surrounds excited by the M-plus stimulus, the S-plus stimulus often had little effect in center or surround (e.g. Figure 2.6A). The S-minus stimuli, though, often elicited strong excitation from the center of these red-on-center cells, a response that was often larger than that to any other stimulus (e.g. Figure 2.6A). Similarly, while the centers of green-on-center cells were suppressed and their surrounds excited by the M-minus stimulus, the S-minus stimulus often produced little response. The S-plus stimulus, on the other hand, often elicited a strong response.

4. A few cells showed opposite responses to the S and M stimuli (e.g. Figure 2.9).

5. Cone-isolating stimuli that employ the same cone fundamentals have been used to study red-green Type I cells in the lateral geniculate (Reid and Shapley, 1992). The S-cone stimulus in these cells was ineffective in driving the cells, and this has been used to argue that geniculate red-green Type I cells do not receive S-cone input. Presumably if the S-cone stimulus significantly modulated the M cones then red-green Type I cells would have responded to it.

6. Other investigators have obtained evidence for S-cone input to red-green cells (e.g. Gouras, 1970; Lennie et al., 1990; Cottaris and De Valois, 1998), and there is even some evidence that the S input aligns with the M input. Vautin and Dow (1985) found that "green" cells were the only ones whose spectral tuning was not matched by the expected (i.e. M) cone fundamental. The spectral tuning included some shorter wavelengths. This could be reconciled by acknowledging the S-cone input to green-on cells, and may also explain why some investigators found blue-green light better than green light for driving color cells (Livingstone and Hubel, 1984).

7. Finally, despite the similarity in appearance between the S-minus and the M-plus stimuli (both look greenish), they elicited opposite responses in most cells.

The use in monkeys of cone fundamentals based on human color matching functions is standard practice, and seems justified (see beginning of Methods). But because of the variability between and within species, it still remains possible that the use of these cone fundamentals is inappropriate, and results in poor cone isolation. This is unfortunately a problem faced by almost all contemporary studies of monkey color physiology because almost all stimuli (both cone-isolating and DKL) use these fundamentals (e.g. Lennie et al., 1990; Reid and Shapley, 1992; Kiper et al., 1997; Cottaris and De Valois, 1998; Seidemann et al., 1999; Chichilnisky and Baylor, 1999).

Though the above discussion suggests the S cone input is real, it is important to note that red-green cells having no S cone input might respond to an S cone stimulus anyway. This is because the S cone stimulus might be subject to longitudinal (or axial) chromatic aberration and consequently might not be cone-isolating (see Note concerning S cone stimulus, Chapter

2§2; but also see Cottaris et. al, 2000). If chromatic aberration is significant then red-green cells having no S cone input would be expected to respond (at least weakly) to the S cone stimulus in a way predicted by the responses to the L-minus stimulus. In fact, the red-green cells did tend to respond to the S-plus stimulus, and these responses did align in space and sign with those to the L-minus responses, consistent with two interpretations: either the cells have significant S-cone input, which happens to align in most cells with the M-plus input (L-minus responses overlap with M-plus responses, Figure 2.12D) or red-green cells do not get S-cone input but respond to the S-cone stimulus because the S-cone stimulus is not actually cone-isolating.

Thus the issue of S cone input into most red-green cells seems unresolved, and likely irresolvable using spatially structured stimuli and silent substitution. Studies using full-field stimuli (in which longitudinal chromatic aberration is not a problem) nevertheless suggest that at least some red-green cells receive significant S cone input (Lennie et al, 1990). In the present study I also found that a very large S-cone stimulus elicited a strong response (data not shown). Finally, the apparent alignment of responses to S and M cone-isolating stimuli might not be coincidence: S cones seem to reside preferentially next to M cones in a bed of abundant L cones (Conway, 2000). This might suggest that development places S cones next to M cones to serve as a retinal substrate for a red-cyan chromatic axis.

2§3.6 Cone interactions

In this set of experiments I designed cone-isolating stimuli that enabled me to stimulate two classes of cones simultaneously. This allowed me to test predictions of how the cone inputs are combined; the plots produced are called cone interaction maps. In order to present two cone-isolating stimuli simultaneously it is necessary to present them on a constant adapting (gray) background (see General design above).

Many reports indicate that color selective cells are mainly unoriented (Livingstone and Hubel, 1984; Ts'o and Gilbert, 1988). I also found this to be so. The cells that did show some mild orientation preference [9/49 cells, class B/C in the four point subjective scale, A = most tuned (Livingstone and Hubel, 1984)] reflected this preference in the asymmetric spatial distribution of their cone inputs (see Figure 2.14A). The orientation preference of these cells was usually only revealed using colored stimuli because the cells responded poorly to white bars (see Figure 2.14B). These mildly orientation-selective cells were useful in studying the cone interactions because the subregions could be conveniently stimulated with bars. The use of bars was useful because it enabled more of the receptive field to be stimulated: the L-plus cone-isolating stimulus looked like a pastel red bar on a gray background, the M-plus looked pastel green on the same gray and the S-plus looked pastel lavender on the same gray.

A further advantage of these stimuli is that they could be overlapped (in space) in a meaningful way: the overlap of L-plus and M-plus stimuli, for example, elicited a relative L-cone activity identical to that of the L-plus stimulus and an M-cone activity identical to that of the M-plus stimulus, leaving the S cones unaffected. This stimulus looked yellowish.

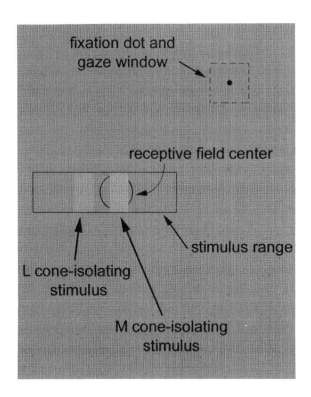

Figure 2.13. Stimulus configuration used to map the cone interactions. The L stimulus was pastel red; the M stimulus was pastel green; the background was gray. Both L and M stimuli were brighter than the background.

I used the reverse correlation two-bar presentation technique (Ohzawa et al., 1997) with eye-position correction (Livingstone and Tsao, 1999) to generate a profile of the response of one cone against another (all plus stimuli). Pairs of bars of optimum orientation were presented simultaneously at random locations along a line perpendicular to the orientation preference of the cell through the receptive field center (Figure 2.14).

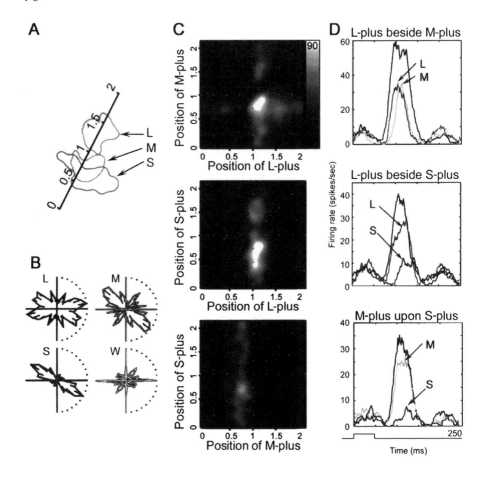

Figure 2.14. Spatial interaction of the cone inputs of an oriented Double-Opponent cell.

A. Spatial response map (i.e receptive field) for the three plus stimuli (L, M, S). The scale bar (in degrees) has an arbitrary origin that forms the axes of the interaction plots (see C).

B. Orientation tuning assessed using bars of cone-isolating light; L-isolating (L), M-isolating (M); S-isolating (S); and white (W). (*continued*)

(*Figure 2.14, continued*)

C. Responses to simultaneously presented pairs of cone-isolating bars. The bars were shifted (in one spatial dimension) along the transect demarcated by the scale bar in (A). Responses for every possible combination of positions that the two bars could have occupied along the transect are given in Cartesian coordinates where the axes of the plots represent the length of this transect. Maximum firing-rate (white) was 90 spikes/second; black is zero spikes/second (not corrected for background).

It is useful to compare the response to a given combination of bar positions (e.g. L-plus bar at location 1.25, M-plus bar at location 0.75) with the position the bars would have occupied in the spatial map given in (A). Note that there was no activity along the x=y diagonal in the top plot (where the two bars would have overlapped) showing that the L-plus and the M-plus bars are mutually suppressive.

D. Post stimulus time histograms corresponding to the peak responses in (C). Medium-gray lines, optimal placement of an L-plus bar; light-gray lines, optimal placement of an M-plus bar; dark-gray lines, optimal placement of an S-plus bar; thick black lines, optimal placement of the two bars together. Stimulus duration indicated at bottom.

The response to simultaneous presentation of these two bars was then plotted in Cartesian coordinates (Figure 2.14C) where at every point in the plot the cell response is given by a gray value (see the firing-rate scale bar to the right of the top panel, Figure 2.14C). The ordinate represents the position of one of the bars; the abscissa represents the position of the other. The diagonal scale bar in Figure 2.14A illustrates the line along which bars of optimum orientation were shifted — this is not the orientation of the bars used to stimulate the cell, but rather the axis perpendicular to the orientation preference. The bars were 1.7 x 0.3° and had an orientation of 67° counter clockwise from vertical. The scale bar in Figure 2.14A is the ordinate and abscissa of the interaction plots (Figure 2.14C). L-plus was mapped against M-plus (top panel, Figure 2.14C), L-plus against S-plus (middle panel, Figure 2.14C) and M-plus against S-plus (bottom panel, Figure 2.14C). The dotted x = y diagonal (Figure 2.14C) represents the locations throughout the receptive field where the bars overlapped. Regions flanking this diagonal represent locations where the bars were adjacent. In interpreting the interaction maps it is useful to relate the response of the cell to the position that the pair of bars would have occupied in the receptive field given in Figure 2.14A. For example, this cell was excited by presentation of an M-plus bar in the region between 0.5 and 1 degree, which is represented in both Figure 2.14A and Figure 2.14C, top panel y-axis.

The response of the cell depended not only on the location of the presentation of one bar but also on the location of presentation of the other. For example, the excitation produced when the L-plus bar was presented in the L-plus-on subregion was cancelled when the M-plus bar overlapped the L-plus bar (Cartesian coordinates of (1.25,1.25) in Figure 2.14C top panel). Likewise, the excitation produced when the M-plus bar was presented to the M-plus-on subregion was cancelled by an overlapping L-plus bar (Cartesian coordinates of (0.75,0.75). Figure 2.14C top panel). This mutual suppression is reflected in the lack of response across the portion of the x = y diagonal that passes through both subregions (from 0.5 to 1.5°, Figure 2.14C, top). In addition, the response to simultaneous presentation of L-plus bars in the L-plus-on subregion and M-plus bars in the M-plus-on subregion was greater than the response to presentation of either bar alone (Cartesian coordinates of (1.25,0.75) in Figure 2.14C, top). Likewise, S-plus bars adjacent to L-plus bars (Figure 2.14C, middle) and M-plus bars on top of S-plus bars (Figure 2.14C, bottom) resulted in increased firing. This is summarized by the post stimulus time histograms (Figure 2.14D).

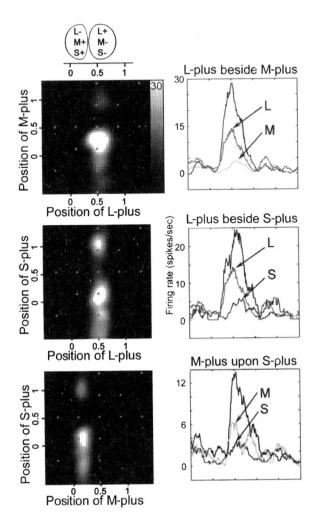

Figure 2.15. Spatial cone interactions for a second cell. Left panels, peak firing-rate (white), 30 spikes/sec. (see Figure 2.14 for details).

The cone interactions for a second cell are shown in Figure 2.15. A diagram of the spatial organization of the receptive field (top of Figure 2.15) indicates the M/S-plus region at 0.25 degrees and the L-plus-on subregion at 0.5 degrees along the stimulus range. This stimulus range forms the axes of the interaction plots (Figure 2.15, left panels). As for the cell shown in Figure 2.14C, the response to simultaneous stimulation of both subregions — e.g., using L-plus bars in the L-plus-on subregion and M-plus bars in the M-plus-on subregion (Figure 2.15 top panel, coordinates of (0.5, 0.25)) — was greater than the response to stimulation of a single subregion (summarized in PSTHs, Figure 2.15, right panels).

Cone interactions were tested for seven cells. In some cells, the peak response to simultaneous stimulation of both subregions was predicted by the sum of the peak responses to separate stimulation of each subregion (e.g. Figure 2.14D); and in others, the response to simultaneous stimulation was greater (e.g. Figure 2.15). The facilitated response, or expansive non-linearity, of some of the cells (e.g. Figure 2.15) is consistent with a simple thresholding operation. Further analysis of a greater sample of cells will be necessary to determine what fraction of cells respond to simultaneous stimulation in a linear way and what fraction of cells respond with an expansive non-linearity, though a different analysis of these cells shows them to be linear (Figure 3.7).

2§4 DISCUSSION

2§4.1 Color constancy and Double-Opponent cells

The paradox of color perception is this: despite varying illumination conditions, the colors we assign to objects are remarkably constant. A red apple, for example, looks red under a blue sky, a cloudy sky and a fluorescent light despite the fact that the spectral distribution of light reflected from the apple are grossly different under each condition. The present study was undertaken because, though it has been shown that

Double-Opponent cells could underlie this color constancy (see Chapter 1), the existence of Double-Opponent cells in the primate has been disputed.

In the present study, I investigated the cone inputs to V-1 cells screened for overt opponency to different colors, cells I call color cells. The combination of cone-isolating stimuli, alert animals, and an eye-position corrected reverse correlation technique enabled a direct assessment of the spatial structure of the receptive fields of these cells, and most were found to be Double-Opponent. There was no evidence that these cells conform to the specifications of "Modified Type II". The surround strengths varied among cells, with some cells showing very weak (and perhaps insignificant) surround responses and others showing surround responses as strong as their center responses; it will be interesting to see what implications this has on the potential for Double-Opponent cells to perform illumination-independent responses. Are there enough of such cells to account for the color constancy of human perception?

2§4.2 The cortical chromatic axes

It has been proposed that parvocellular geniculate cells, the majority of which are excited by L cones and suppressed by M cones, or vice versa (Wiesel and Hubel, 1966, Derrington et al, 1984, Reid and Shapley, 1992), provide a substrate for a red-green chromatic axis. This axis is thought to work with a blue-yellow axis, represented by the blue-yellow retinal bistratified cells (Dacey and Lee, 1994) and geniculate Type II cells (Wiesel and Hubel, 1966), and a black-white axis to represent color space (Livingstone and Hubel, 1984). The situation is not so simple in V-1 because distinct chromatic axes are not thought to exist (Lennie et al, 1990). Rather, cortical cells are thought to be tuned to many different colors (Lennie et al, 1990; Cottaris and De Valois, 1998).

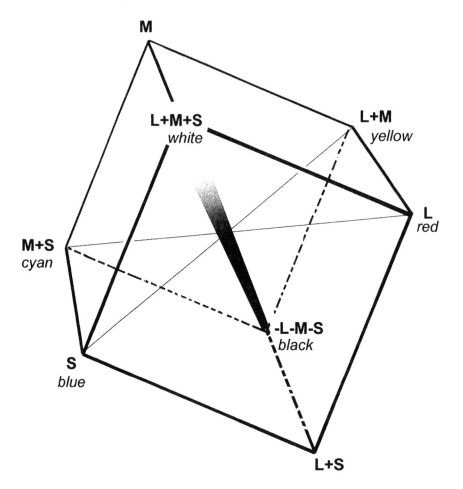

Figure 2.16. The chromatic axes may be described loosely as a cube, where the corners of the cube represent maximal cone activity. The cube encompasses all colors; each color is described by a unique set of three values; each value represents the point along one of the axes through the cube. Many red-green (i.e. L vs. M) cells responded to the S-cone stimulus. Could the red-green axis actually be red-cyan?

In fact it has been argued that single cortical cells "multiplex" color selectivity and other visual attributes including orientation selectivity (Lennie, 2000). The majority of cortical cells are orientation selective; moreover, the optimal color of a stimulus will be different for each cell because the strength of the input from each cone class varies among cells (Lennie et. al, 1990; Cottaris and De Valois, 1998). But any such "color selectivity" need not imply that these cells are responsible for color perception: the different weights of the cone inputs could reflect the fact that cortical cells are sampling a relatively small number of cones from a patchy cone mosaic, and that evolution has not had to go to the trouble of guaranteeing that the non-color cortical cells have identical cone inputs. It seems more likely that the cortical color-coding cells are the relatively rare cells that show explicit opponency between cone classes. After all, such opponency is the hallmark of color perception (Hering, 1964). Thus in the present work I only studied cells that showed overt opponency between cone classes. (I further restricted my study to L vs. M cells, i.e. red-green cells.)

Red-on-center cells were suppressed by M cone-isolating stimuli to the same extent that they were excited by L cone-isolating stimuli; similarly green-on-center cells were suppressed by L cone-isolating stimuli as well as they were excited by M cone-isolating stimuli (Figure 2.12A). This shows that these cells are concerned with the color and not the luminance of the stimulus: red-on-center cells, for example, responded equally well to a stimulus that appeared bright red (L-plus) and to a stimulus that appeared dark red (M-minus). Moreover, the fixed ratio of L and M-cone inputs (Figure 2.12A,B) argues that these cells are encoding a single chromatic axis. This axis would presumably be complemented by the well-documented S vs (L+M) (or blue-yellow) axis (Dacey and Lee, 1994) and a luminance (or black-white) axis. Three axes are sufficient to describe all of color space (Figure 2.16).

Specific hues are likely encoded by the cumulative activity of cells representing these three axes. In the same way a specific hue on a computer monitor requires values for all three phosphors (R, G or B), the perception of specific hues would require the activity of all three chromatic axes. Moreover, only three cells (one for each chromatic axis) would be required for every receptive field location. This seems an efficient means of encoding color, and is consistent with the number of color cells observed here (~10% of cortical cells were red-green; 7 times fewer cortical cells are blue-yellow (Livingstone and Hubel, 1984)).

Some color cells respond to luminance (e.g. Figure 2.6A and Johnson et al, 2001), which, taken by itself, has been used to suggest that these cells multiplex color and luminance information (Johnson et al, 2001). Color cells would not respond to luminance if the opponent cone inputs were perfectly balanced (a luminance stimulus is L-plus superimposed on M-plus). Thus, that these cells respond to luminance shows that the opponent cone inputs are not perfectly balanced (hence the non-unity r^2 value in Figure 2.12A), but it need not be the case that these cells use this imbalance to relay information about luminance. Rather, the imbalance might not be biologically significant, and might have arisen because perfectly balanced cone inputs does not provide a significant natural-selective advantage. After all, the population as a whole shows balanced opponent inputs (hence the slope of ~1 in Figure 2.12A) — and presumably it is the population that encodes information not single cells.

Cortical red-green cells usually responded to the S-cone stimulus (Figure 2.12D). This response may be due to a compromise of cone isolation attributed to chromatic aberration (see Note concerning S cone input, Chapter 2§2 above) but it also may reflect genuine S cone input. If this S-cone input is real, the cortical red-green axis may be better described as red-cyan (or L vs. ½(M+S), as suggested by Figure 2.12, since responses to the S stimulus were usually aligned in space and sign with those to the M stimulus (in 93% of cells). A red-cyan axis might be advantageous because it (and the blue-yellow axis) would be silent to shades of gray (Figure 2.16). That is, the intersection of the null planes of the two axes would be achromatic. This is not true for the conventional red-green (L vs. M) and blue-yellow axes.

If the response to the S-cone stimulus reflects an S contribution, some may find it surprising that it aligns with the M input and not the L input. Long-wavelength light activating the L cones appears reddish, but so does very short-wavelength light activating the S cones (Ingling, 1977). Some have argued that the "reddish" quality of short-wavelength light implies a contribution of the S cones to the red-green axis, aligning with the L cones (Ingling, 1977), but others have shown that this need not be the case (Krauskopf et al, 1982). In fact, an independent component analysis of color in natural images has shown that the best way to represent color information is with oriented double-opponent blue-yellow, red-cyan and black-white cells (Tailor et al, 2000; Wachtler et al, 2001).

2§4.3 Wiring Double-Opponent cells

Most parvocellular geniculate Type I cells are color opponent (Wiesel and Hubel, 1966; De Valois et al., 1966; Reid and Shapley, 1992) (Figure 1.5) and one might assume they supply cortical Double-Opponent cells. On the other hand, Type I cells (at least those subserving the fovea) might be color opponent as a byproduct of their high spatial resolution (Lennie, 1980; Calkins and Sterling, 1999), and consequently the involvement of Type I cells in color perception (and in wiring Double-Opponent cells) might be an unjustified assumption. For this reason, Type II cells — the receptive fields of which

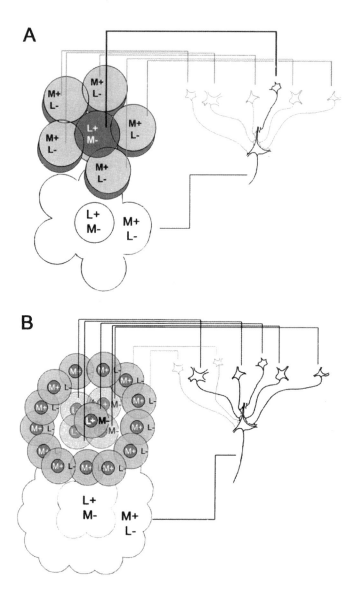

Figure 2.17. Models describing the hypothesized inputs into a Double-Opponent cell. See text for details.

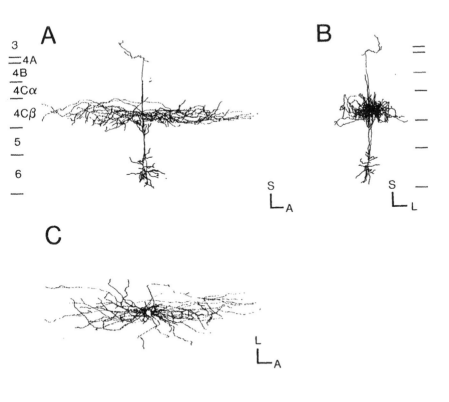

Figure 2.18. Morphology of a color cell (from Figure 4 of Anderson et al. (1993) *Cerebral Cortex* **3**, 412-20). Parasagittal view (A), transverse view (B) and surface view (C).

comprise many cones (Wiesel and Hubel, 1966; Chichilnisky and Baylor, 1999) — may form the dominant input into cortical Double-Opponent cells (Hubel and Livingstone, 1990; Rodieck, 1991). This agrees with the similarity in size of Type II cell receptive fields and the centers of cortical Double-Opponent cells. Moreover, the LGN interlaminar leaflets (koniocellular layers), which contain Type II cells (Livingstone and Hubel, 1984; Martin et al., 1997), project to the cytochrome oxidase rich blobs (Livingstone and Hubel, 1982; Diamond et al., 1985), regions that are rich in Double-Opponent cells (Livingstone and Hubel, 1984; Ts'o and Gilbert, 1988).

In this scheme, a small number of Type II cells (possibly only one) would feed the center of a Double-Opponent cell, while several Type II cells of opposite chromatic tuning would feed the surround (Figure 2.17A; Michael, 1978). A major shortcoming of this hypothesis, however, is that geniculate red-green Type II cells have not been documented —besides the original assertion of their existence (Wiesel and Hubel, 1966; Livingstone and Hubel, 1984).

Thus parvocellular Type I cells might be the only available substrate for cortical red-green Double-Opponent cells (Figure 2.17B). In this model, the center could be formed by pooling geniculate Type I cells of similar chromatic tuning; the surround could be formed by intracortical lateral connections known to exist between patches of cortex that are rich in color cells (Livingstone and Hubel, 1984). Thus Michael's model (Figure 2.17A) might be modified slightly so that the substrate for Double-Opponent cells would be cortical (and not geniculate) red-green Type II cells. That geniculate Type I cells are the substrate for cortical red-green cells is supported by two findings: first, the morphological description of one color cell showed that it arborized heavily in layer $4C\beta$ (Figure 2.18; Anderson et al., 1993; Layer $4C\beta$ is the cortical target of parvocellular lateral geniculate cells and itself is thought to contain Type I cells (Livingstone and Hubel, 1984).) And second, Type I cells have well balanced L vs. M inputs (Derrington et al, 1984), which agrees with the balanced cone inputs seen in cortical red-green cells (Figure 2.12A). Finally, Type I cells in the peripheral retina, where the cells' receptive-field centers are fed by many cones, are chromatically opponent (Martin et al, 2001), supporting the idea that Type I cells do not acquire their color selectivity by accident.

In sum, most V-1 cells probably do not multiplex multiple visual attributes (e.g. form and color). The cells I have studied, for example, have large receptive fields and lack sharp orientation selectivity, making them an unlikely substrate for high-resolution form perception. Instead, they have receptive fields that are both spatially and chromatically opponent making them well suited to subserve color perception. The large receptive field sizes of Double-Opponent cells might limit the spatial resolution they could encode. Indeed, this is consistent with our lower acuity for images in which color is the only cue (Liebmann, 1926; Granger and Heurtley, 1973; De Valois and Switkes, 1983; Mullen, 1985; for review see Livingstone and Hubel, 1987).

The responsiveness of many well-oriented cells to color (Hubel and Wiesel, 1968; Livingstone and Hubel, 1984) — and the fact that these cells respond to red bars on equaluminant green backgrounds (Gouras and Kruger, 1979; Thorell et al, 1984; Hubel and Livingstone, 1990; Johnson et al, 2001) — should not be taken as evidence that these cells "encode color". Rather, for a cell to be capable of encoding hue (a "color cell"), it would also need to show explicitly opponent responses to opponent colors (e.g. excitation to red and inhibition to green). After all, it is this sort of opponency that is the hallmark of color perception (Hering, 1964; Hurvich, 1981).

Finally, the fixed ratio of L and M input that red-green Double-Opponent cells receive (Figure 2.12) suggests that these cells establish a single chromatic axis which, in conjunction with a blue-yellow and a black-white axis, is sufficient to describe all of color space.

Chapter 3

Temporal structure of cone inputs to cortical color cells

"The pure spectral colors, which are the most saturated objective colors in existence, still do not elicit in the unfatigued eye the most saturated color sensation that it is possible to have. This highest degree of saturation is obtained by making the eye insensitive to the complementary colors"
— Helmholtz, 1909

3§1 SUMMARY AND CONCLUSIONS

The color of a stimulus can be made more salient if it is immediately preceded by a stimulus of opposite color, a phenomenon known as successive or temporal color contrast (Helmholtz, 1909, pg. 243; Figure 3.1). Here I tested whether color cells in monkey V-1 could mediate this temporal chromatic contrast. I did this using sequential colored stimuli (consisting of small patches of cone isolating light) presented to the centers of V-1 cell receptive fields. All colored stimuli were brighter than the constant adapting background on which they were presented. For almost all color cells (33/35), the response to a stimulus was reduced if the stimulus was immediately preceded by a similarly colored stimulus. But the response was increased if the stimulus was preceded by an oppositely colored stimulus. I tested 12 non-color cells in a similar way; for all of these cells, I found that a previous stimulus, regardless of its color, reduced the response to a subsequent one. Thus color cells are specifically capable of encoding temporal color contrast. Many color cells were also Double-Opponent, a receptive field organization that makes them well suited to code spatial color contrast. Thus these cells seem well equipped to mediate two key features of color vision: simultaneous and successive color contrast.

Cells in higher visual areas (V-4) thought to be critical for color vision (Zeki, 1983) have a color preference but generally do not show overt color opponency within a restricted part of visual space (Schein and Desimone, 1990); i.e. V-4 cells do not generally show off-excitation or on-inhibition. It would seem that cells would need to show off-excitation or on-inhibition if they were the site of the mechanism for temporal chromatic contrast. It is tempting to speculate that color percepts are therefore generated in V-1: color percepts could not be generated before V-1 because Double-Opponent cells (which are necessary for spatial color contrast) do not arise until V-1; and, it seems unlikely that color percepts are generated after V-1 because the necessary color opponency required for temporal color contrast is lost after V-1 (at least in V-4). An alternative interpretation is that a different extrastriate area (not V-4) is responsible for generating temporal color contrast and hence color percepts. One study in humans has identified one such area: V-8 (Hadjikhani et al, 1998). This area does not seem to correspond to area V-4 in monkeys. So the obvious question remains, is there a monkey homologue of V-8?

3§2 METHODS

3§2.1 General Design

The general design of this experiment is identical to that given in Chapter 2§2, in all regards except two. First, though the stimuli were cone-isolating (as in Chapter 2), all were presented on a constant gray adapting background (low cone-contrast stimuli), rather than differing adapting backgrounds (high cone-contrast stimuli). (For a discussion of cone-isolating stimuli and the nomenclature of the stimuli see Chapter 2§2). The main object of this study was to measure responses of cortical cells to successive differently colored stimuli. The backgrounds had to be constant among stimuli so that any effect on the response of a neuron to the successive stimuli could be attributed to a change in color of the stimuli and not to a change in color of the background or a change in mean luminance. The high cone-contrast stimuli are not useful for this purpose because all the stimuli have different adapting backgrounds and different mean luminances. Second, in this experiment I quantified the responses of some non-color cells; i.e. some cells that did not show overt opponency to the L and M-plus stimuli.

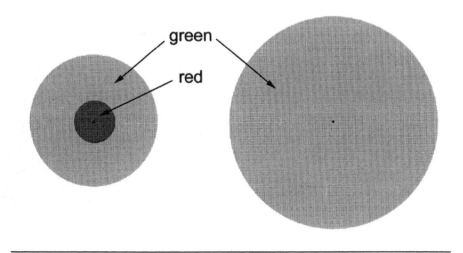

Figure 3.1. Demonstration of successive temporal color contrast. (*continued*)

(*Figure 3.1, continued*) "When the primary color is complementary to the reacting color of the ground, the latter color comes out more saturated in the extension of the after-image than it does on the parts of the retina that have not been fatigued or on the parts that have been fatigued by the color of the ground" (Helmholtz, 1909, pg. 243). Draw two bluish-green discs of slightly different sizes, as above, placing a small red disc in the center of the smaller. Then stare at the fixation dot in the center of the red disc. After 20 seconds or so, transfer your gaze to the fixation dot in the center of the large bluish-green disc. You should see, as Helmholtz described, a supersaturated color corresponding to the sum of the afterimage to the red disc and the image of the bluish-green ground. This afterimage is heightened further because the afterimage surrounding that of the red disc (i.e. the afterimage to the bluish-green ground) is red which will mix with the bluish-green image to form a desaturated percept.

3§2.2 Stimulus presentation

Temporal interactions were studied for color cells with receptive field centers larger than 0.3 degrees; i.e. not Type I cells. To determine the temporal interactions, only plus stimuli were used. (This is in distinction to the generation of the spatial maps where both plus and minus stimuli were used, see Chapter 2§3) My choice to use the plus stimuli instead of the minus stimuli was arbitrary, but it was necessary to pick either all plus or all minus stimuli because all the stimuli had to be similarly bright. The minus stimuli have a lower mean luminance than the gray background and the plus stimuli have a higher mean luminance than the background. It was important to make sure all the stimuli were similarly bright to ensure that the cells were not responding to a change in the luminance of the stimuli.

All stimuli were presented on the same gray background: the L-cone isolating stimulus appeared pastel-red; the M-cone solating stimulus appeared pastel green; and, the S-cone stimulus appeared pastel-lavender. The L-cone contrast was L_L-L_G/L_L+L_G), where L_L was the cone excitation of the L-plus stimulus and L_G was the cone excitation due to the gray background. Cone excitations were calculated by taking the dot product of the emission spectra of each stimulus [measured using a Photo Research PR 650 SpectraScan spectrophotometer Chatsworth, Calif.)] and the three cone absorption spectra Smith and Pokorny, 1972, 1975). The L and M cone contrasts were both 0.21, and the S cone contrast was 0.49.

There is some debate about the validity of the Smith and Pokorny cone fundamentals, especially that of the S cone. However under the recording conditions used here, results using cone-isolating stimuli based on more recent fundamentals ((Stockman and Sharpe, 2000; Stockman et al., 1999) are comparable to those using stimuli based on Smith and Pokorny fundamentals (see Chapter 2§2). The reader is directed to Chapter 2, *Note concerning the S cone stimulus* for a discussion of the possible effects of macular pigmentation and chromatic aberration on cone-isolation.

3§3 RESULTS

3§3.1 Distinguishing color from non-color cells

In Chapter 2§3 I introduced a technique for mapping the spatial cone interactions of color cells. However, I was curious what the spatial cone interactions were for non-color cells; i.e. for cells that did not give opponent responses to the L-plus and M-plus high-cone contrast stimuli. As I will show, the interactions between L and M-plus stimuli turn out to be a rather effective means of confirming the distinction between color cells and non-color cells.

Around 615 cortical cells were screened with cone isolating stimuli: every cortical cell encountered was tested with alternating flashes of a small patch (less than 1 x 1°) of L plus and M-plus cone-isolating light centered on the receptive field (see Figure 2.3 and 2.5). The patch was manually guided in and out of the receptive field to find the cell center. (Cells were also tested with spots and oriented flashed and moving bars of various colors.) 73 cells were designated as color cells because they responded vigorously to small spots of colored light and the L-plus and the M-plus stimuli produced opposite responses (excitation vs. suppression) (Conway, 2001). (8 of the 73 were not studied here because they were classified as Type cells, see 2§3.2).

The method for mapping the cone interactions is identical to that presented in Chapter 2§3 (Figure 2.14). Pairs of cone isolating stimuli were presented simultaneously at random locations in one spatial dimension along a stimulus range centered on the cell's receptive field. For a given cell, the stimulus range was much larger than the receptive field and each stimulus was the same size or smaller than the receptive field center. Stimuli remained on the monitor for 48-200 ms there was a minimal refresh delay (<13 ms) between stimuli presentations. The location of each stimulus in the stimuli pairs was independently generated and the stimuli were free to overlap. When the stimuli overlapped, they did so in a meaningful way. For example, the overlap of L-plus and M plus stimuli elicited a relative L-cone activity identical to that of the L-plus stimulus and an M-cone activity identical to that of the M-plus stimulus, leaving the S cones unaffected. This stimulus appeared yellowish. A continuous record of eye position, stimulus position and spike history was collected.

I determined the average response that the cell gave to every possible spatial configuration of the two stimuli (along the stimulus range) using the two-bar presentation technique (Ohzawa et al., 1997) with eye-position correction (Livingstone and Tsao, 1999) and corrected for the visual latency. In the interaction plots (Figure 3.2) the ordinate represents the position of the L-plus bar along the stimulus range; the abscissa represents the position of the M-plus bar along the stimulus range. The diagonally displaced dots in the plots are placed along the same-position diagonal, and represent the locations throughout the receptive field where the stimuli overlapped. The stimulus range was centered on the receptive field; note that the stimulus range has an arbitrary origin (i.e. this is not the eccentricity of the cell in visual space). The response of the neuron for every combination of bar positions is given by a color scale bar (Figure 3.2A). To make sense of these maps it is useful to relate the spatial configuration of the two bars for a given set of coordinates in the interaction plot with the spatial location that the two bars would have occupied in the receptive field (a diagram of the receptive field and the stimulus range is given for reference above the interaction plots, Figure 3.2).

As for all the color cells for which cone interactions were determined, there is little or no activity along the same-position diagonal: a stimulus of superimposed L- and M-plus bars is not effective anywhere in the receptive field. This is not surprising, given the result reported in Chapter 2§3 (Figure 2.12 A, B): color cells are chromatically opponent in every receptive field subregion (i.e. both center and surround).

In contrast, non-color cells did show activity along the same-position diagonal (Figure 3.2B). In fact this activity was often greater than the activity to either stimulus alone. This result is perhaps not surprising given "non-color" cells likely care only about the relative brightness of the stimulus.

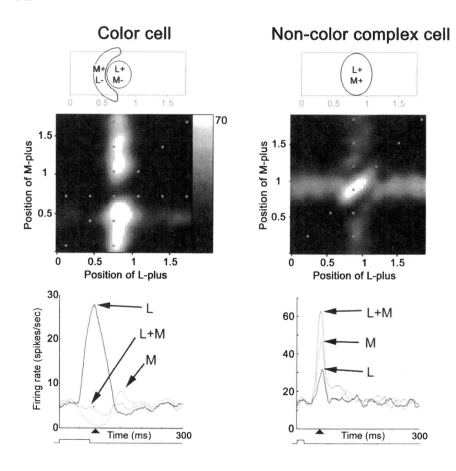

Figure 3.2. Spatial cone interactions.
Top panels. On the abscissa is plotted the response to the L-plus stimulus and on the ordinate is plotted the response to the M-plus stimulus. White represents maximum firing; black represents a firing-rate of zero spikes/second. The x = y diagonal represents locations where the L- and M-plus stimuli overlapped (such a stimulus looks yellowish). The stimulus range is shown over a diagram of the receptive field at top. (*continued*)

Figure 3.2 continued) **Bottom panels**. The response (PSTH) to stimuli presentation to the center of the receptive field; L-plus bars, dark-gray lines; M-plus bars, light-gray lines; overlap of M- and L-plus bars, medium-gray lines.

A stimulus composed of overlapping M- and L-plus bars is brighter than either an L- or M-plus bar alone, thus one would expect a simple luminance detector to respond better to it.

For non-color cells, diagonals flanking the same-position diagonal often showed a decrease in activity — these are the two darker bands flanking the orange diagonal, centered on the same-position diagonal. This shows that a pair of stimuli presented simultaneously and adjacent to each other, regardless of color, was less effective than a single stimulus presented alone. This is easily interpreted if one appreciates that these cells are complex. Complex cells have a relatively large receptive field over which they respond best to oriented bars; but if the bars are made wide to encompass the entire region over which thin bars are effective the response is attenuated (Hubel, 1962; Movshon et al, 1978). To a cell that cares only about the luminance of a stimulus and not its color, pairs of adjacent L-plus and M-plus bars would simply look like a single wide bar, and be a less effective stimulus.

In contrast, many color cells showed an increase in activity along diagonals flanking the same-position diagonal (in the example shown in Figure 3.2A the increase is clear only in one of the flanking diagonals). This shows that these cells respond best to adjacent differently colored bars, and is a result not unexpected given these cells are often Double-Opponent (Chapter 2§3).

The striking difference in the cone-interaction maps between color and non-color cells reinforces the notion that

color and non-color cells represent two distinct populations o
cells, responsible for encoding different aspects of the stimulus.

3§3.2 Temporal structure of cone inputs

From the same spike train used to map the spatial cone
interactions I also generated space-time maps of the response to
each of the stimuli (L-plus, M-plus, and sometimes S-plus
(Figure 3.3). The abscissa is the stimulus range (see Figure 2.14
along which the stimuli were shifted; the ordinate is the delay
before each spike, with a delay of zero at the bottom of each
map.

While the monkey fixated, stimuli were presented at 37 or
75 Hz along at random positions along a stimulus range
running through the center of the cell's receptive field (Figure
3.2, top). After presenting several thousand stimuli, I plotted
the average response to stimuli presentations at each point
across the receptive field, accounting for eye position. The
maps (Figure 3.3) can be thought of as a collection of post
stimulus time histograms for stimulation at each point across
the receptive field, with the onset of the stimuli aligned along
the abscissa, and the level of activity indicated according to the
gray-scale firing-rate bar. White represents high firing; black
represents zero firing.

At the visual latency of a given cell (~ 50 ms) , the pattern
of activity across the stimulus range reflects the receptive field
at very short delays and very long delays the activity is
random. Randomness is reflected as homogenous bands
throughout the receptive field, the temporal frequencies of
which are determined by the stimulus durations (indicated at
the left, Figure 3.3).

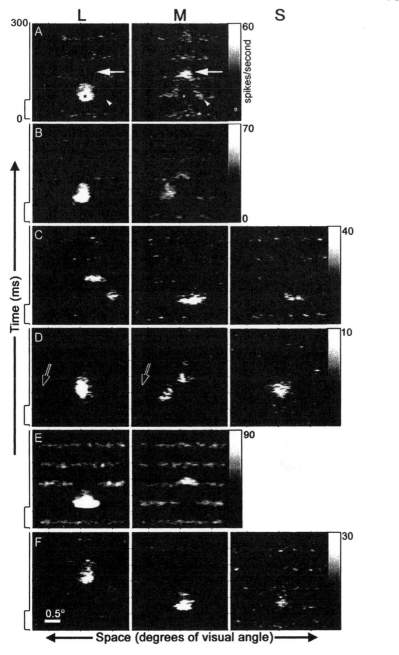

Figure 3.3. Space-time maps of six color cells. (*continued*)

(Figure 3.3 continued). Space-time maps of six cortical color cells (**A-F**) to cone-isolating stimuli (L, M or S). For each plot, the abscissa represents the positions in space (the stimulus range along which bars of cone-isolating light were shifted. The ordinate is time, with stimulus onset aligned along the bottom of each map; the stimulus duration is indicated at left. The firing-rate is given by a gray-scale rate where white represents maximum firing and black, zero. Note that in many cells the response for a given receptive field subregion is biphasic. For example, the cell in (**A**) was initially excited by an L-plus stimulus (asterisk) and subsequently suppressed by it (arrow) the cell gave the opposite response pattern to the M-plus stimulus. In some cells (**A, B, C, D**) a chromatically-opponent surround was clear (arrowhead in A). Scale bar is 0.5°. All cells were 2.5 to 5° away from the fovea.

A dark region represents suppression below baseline if it is darker than the average gray value outside the receptive field for a given delay (for example the asterisk in the M map and arrow in the L map, Figure 3.3A); similarly, white represents excitation if it is above baseline for a given delay (e.g. asterisk in the L map, Figure 3.3A). The firing-rate has been normalized taking into account the number of bar presentations. All the cells were chromatically opponent; for example, in Figure 3.3A at short delays (asterisks), spikes were often elicited by placement of an L-plus bar in the receptive-field center but rarely elicited by an M-plus bar. Many cells were Double Opponent (Figure 3.3A, B, C, D): in the receptive field region surrounding the center the pattern of probability is reversed (e.g. Figure 3.3A, arrowhead).

In all experiments I used stimuli that were well matched to the size of the receptive field center. In a few cases, however, also used stimuli that were much wider than the receptive field. This seemed like a fun way to explore further the Double Opponency of these cells. The rationale was simple: if the cells are Double-Opponent then one would expect them to respond preferentially to the edges of a colored stimulus. By making the edges further apart (for example by making the stimulus wider) it should be easy to see such a response in the space-time maps. In fact, the cells did respond better to the edges (Figure 3.4), though using wide stimuli seemed to kill the biphasic component of the response.

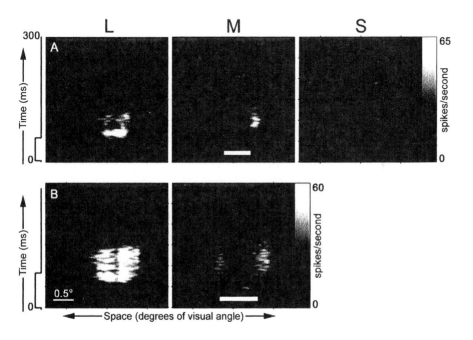

Figure 3.4. Space-time map of two L-plus centered color cells tested with stimuli that were wider than the receptive field center. Conventions as for Figure 3.3. Note that the activity is highest in the region corresponding to the edges of the stimulus: the cells responded to the inside edges of the L-plus stimulus and the outside edges of the M-plus stimulus. The white bar in the M-plus map gives the width of the stimulus used for that cell.

3§3.3 Interpreting space-time maps

At very short delays and very long delays, the probability that a stimulus elicited a spike is random across the receptive field. This is reflected in the space-time maps as horizontal gray bands, which represent the average activity of the cell, disregarding the spatial location of the stimulus. The spatial structure of the space-time maps only reflects the spatial structure of the receptive field at time delays that correspond to the visual latency of the cell. Different aspects of a cell's response can have different latencies; e.g. the cell can be both suppressed and excited by the same stimulus in the same receptive field subregion (Figure 3.3) but with different latencies. Space-time maps are a nice way of showing this aspect of the cells' responses.

But one must be careful in interpreting space-time maps. When a bar is presented outside the receptive field of a cell, the cell is unlikely to fire. This low probability of firing only holds for delays corresponding to the visual latency of the cell and shows up in space-time maps as black regions adjacent to the receptive field (e.g. open arrow, Figure 3.3D). However, suppression would also show up as black regions. In order for a given black region to be interpreted as suppression, it must be blacker than the average color outside the receptive field, for a given delay (for example, the center of the receptive field at short delays, Figure 3.3B, M-plus map).

3§3.4 Responses to sequential colors

Many cells showed a biphasic response in time: the probability that a bar preceded an action potential switched from high to low, or vice versa (e.g. Figure 3.3A, from asterisks to arrows).

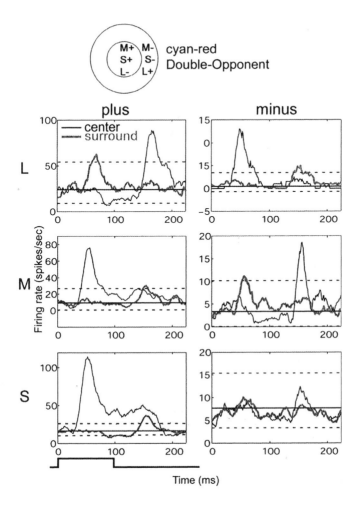

Figure 3.5. Temporal structure of a red-cyan Double-Opponent cell. Segments of the reverse correlation data corresponding to stimulus presentation in the center (black traces) and surround (gray traces) for the plus stimuli (left panels) and minus stimuli (right panels) were collected for a Double-Opponent cell in which the spatial structure was assessed using high-cone-contrast stimuli (see Chapter 2§3). Stimulus duration indicated at bottom.

100

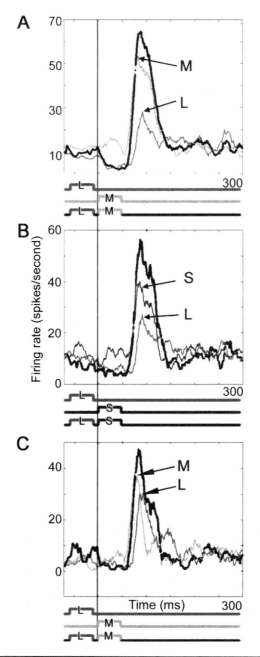

Figure 3.6. Responses of color cells to sequences of colors (*con't*)

(*Figure 3.6 continued*) Responses of color cells to their preferred (i.e. reference) stimulus are increased if the stimulus is preceded by an oppositely colored stimulus. The response PSTHs of three cells; a green-on cell (**A**), a blue-on cell (**B**) and a red-on cell (**C**). The responses to single stimuli (preceded by the constant adapting gray background) are given by the colored lines. Green, M-plus stimuli; red, L-plus stimuli; blue, S-plus stimuli. The responses to sequences of stimuli (duration and color indicated at bottom of each PSTH) are given by the black traces. The vertical line indicates the onset of the reference stimulus; e.g. for the green-on cell this is the M-plus bar.

In fact this was true for both center and surround (for those cells having substantial surround) as shown in Figure 3.5. The responses to center (black lines, Figure 3.5) and surround (gray lines, Figure 3.5) were extracted from the reverse correlation data for a cell for which spatial maps were collected using high-cone-contrast stimuli (see Chapter 2 §3).

The biphasic nature of the response suggests that these cells might respond differently to sequences of colored stimuli than to single stimuli. That is, a preceding stimulus might modify the response to a subsequent one, depending on the colors of the stimuli and the type of cell. Such a pattern of response could be responsible for temporal color contrast. In the next set of experiments I tested this explicitly (Figure 3.6). For each cell, the stimulus that produced excitation with the shortest latency (in the center of the receptive field) was designated the reference stimulus. Thus for the cell shown in Figure 3.6A, the reference stimulus was the L-plus stimulus. Responses to sequential stimuli were only examined for stimuli presentations to the center of the receptive field.

Most color cells showed an increase in response to the reference stimulus if the stimulus was preceded by a stimulus of opposite color contrast (Figure 3.6). In most cases the response to the preferred sequence was closely approximated by the

102

linear sum of the response to the reference stimulus alone plus the off-response to the preceding oppositely colored stimulus (Figure 3.6A and B, but see Figure 3.6C).

As quantified in Figure 3.7, the expected value (the sum of the on-response to the reference stimulus and the off-response to the opposite color stimulus) predicted the response to the observed value (the response to the sequence) (slope, 0.95 $r^2=0.8$).

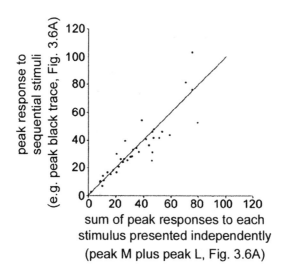

Figure 3.7. The response to sequences of colored stimuli (preferred color preceded by non-preferred color, ordinate) is predicted by the sum of the on-response to the preferred color plus the off-response to the non-preferred color (abscissa). The background firing-rate was subtracted from the measurements. The x = y diagonal is shown for reference.

I also quantified the responses in a different way. Three situations were examined: 1, the response of the cell to the reference stimulus alone (i.e. preceded by the gray background); 2, the response to the reference stimulus preceded by a stimulus of identical color; and 3, the response to the reference stimulus preceded by a stimulus of opposite color (Figure 3.8). The responses to sequential colored stimuli were quantified by calculating the percent change of response as compared to the response to the reference stimulus alone (Figure 3.9).

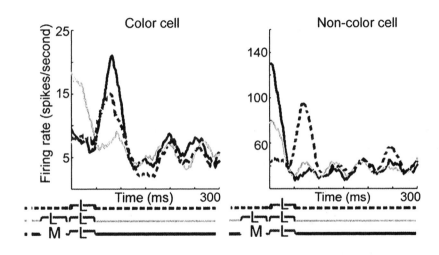

Figure 3.8. Color cells, but not non-color cells, respond better to sequences of oppositely colored stimuli than they do to single stimuli or to similarly colored stimuli. The reference stimulus in for both cells was an L-plus bar. Note that for the color cell the response to this reference (dotted lines) was increased if the stimulus was preceded by an oppositely colored (i.e. M-plus) stimulus (solid lines) but decreased if the stimulus was preceded by a similarly colored stimulus (gray lines). In contrast, for the non-color cell the response to the reference stimulus was always decreased regardless of the color of a preceding stimulus.

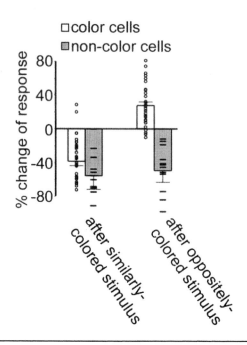

Figure 3.9. Color cells (n=36), but not non-color cells (n=12), responded better to sequences of oppositely colored stimuli than they do to single stimuli or to similarly colored stimuli.

I found that there was variability in terms of the percent change of response, but in general the response to a sequence of oppositely colored stimuli was increased by 28 % (+/- 5% std. error) (Figure 3.9). In contrast, the response of non-color cells was never increased by a preceding stimulus. In fact the response of non-color cells was decreased by any preceding stimulus regardless of color (Figure 3.9), reminiscent of the neuronal correlate of forward masking (Macknik and Livingstone, 1998).

The response of color cells was not uniformly increased by a preceding stimulus: their responses to reference stimuli were decreased if the stimuli were preceded by ones of a similar color

(Figure 3.9). The pattern of responses to sequential stimuli shown by color cells is consistent with the psychophysical observation that a stimulus can be made more salient if it is preceded immediately in time by an oppositely colored stimulus, and less salient if it is preceded by similarly colored stimulus (Figure 3.1). This psychophysical phenomenon is known as temporal color contrast.

3§4 DISCUSSION

3§4.1 Temporal chromatic contrast in macaque V-1

Color cells early in the visual pathway are fed by several cone classes that are pitted against each other: one cone class excites a given color cell and a second cone class or classes suppresses it. For example, the bistratified retinal ganglion cell is excited by S cones and suppressed by some mixture of L and M cones (Dacey and Lee, 1994). Color afterimages are thought to arise when light reflected from a vividly colored surface selectively renders one class of cones ineffective and therefore unresponsive, a process called chromatic adaptation. After viewing such a surface, the cone class(es) that provide the opponent input to a downstream ganglion cell would be left unopposed, thus tipping the balance to produce an illusory color. Color afterimages contribute to temporal color contrast. For example, a green can be made to appear more saturated if a red immediately precedes it (the afterimage of the red sums with the image of the green to produce a supersaturated green).

In this chapter, I asked whether color cells in primary visual cortex might be able to take advantage of their cone-opponency to subserve temporal color contrast. Not surprisingly they are, as shown by the fact that they respond in predictable ways to temporal sequences of colored stimuli. This was not true of the non-color cells studied.

Chromatic adaptation begins in the photoreceptors. With extended exposure to an appropriate strong wavelength, a photopigment will bleach, will no longer be able to absorb light and will cripple the photoreceptor carrying that photopigment (Rushton, 1961). Exposure to bright stimuli results in substantial photopigment bleaching and also produces long-lasting afterimages, the duration of which is consistent with the time required for photopigment recovery.

It is unclear whether photopigments bleach significantly upon exposure to less bright stimuli, of the sort used in the present study, though chromatic adaptation clearly takes place (afterimages to the stimuli used here are vivid, if short lived). Possibly instead of photochemical bleaching (or in addition to it), the adaptation underlying color afterimages to less bright stimuli is neural. Neural adaptation begins in the photoreceptors: calcium influx through cG activated channels (which open in the dark and carry the "dark" current) function in a negative feedback loop to shut off the production of cG, thus closing the cG activated channels (Rodieck, 1998, pg 398-9). Neural adaptation is likely the source of the color-contrast phenomenon described here for another reason: the off-discharge that underlies the temporal color contrast is short-lived, which is typical of neural adaptation [(Rodieck, 1998) but see (Virsu and Laurinen, 1977)] and not so typical of photochemical bleaching (Brindley, 1962).

One question that remains is this: the afterimages produced by the stimuli used here last a long time (minutes) if one is exposed to the stimuli for a reasonable length of time (minutes). It does not appear that the duration of the off-discharge increases in response to longer stimulus exposures (based on preliminary experiments). So, if the off-discharge does not account for the long-lasting color afterimage, what does? Presumably cells in the visual cortex are responding in order to produce the long-lasting color afterimages (Burr et al, 1987). One solution is that chromatic adaptation, and afterimages, have two underlying causes with different time courses (Geisler, 1978). A photochemical bleaching one that is long-lasting and underlies long-lived color afterimages; and a neural one that is short-lasting and underlies the transient saliency of temporal color transitions. The temporal color contrast phenomenon described here is accounted for by the short-lasting neural adaptation. Perhaps by examining the activity of color cells over longer time periods one would gain insight into how the longer-lasting afterimages are encoded.

Chromatic adaptation of the sort discussed above can only be preserved in cells that maintain cone opponency (this is in distinction to post-receptoral adaptation, which takes place at some higher stage of color processing (Krauskopf et al, 1982; Gerling and Spillman, 1987; Webster and Mollon, 1991). In the primary visual cortex, these cells seem to be the Double-Opponent cells[1]. With the exception of some V-2 cells (Hubel and Livingstone, 1987; Gegenfurtner et al, 1996), Double-Opponent cells seem to be the furthest cells from the sensory receptors that still preserve cone-opponency — cells in V-4, the "color area" of the monkey (Zeki, 1983), rarely show overt color opponency (Shein an Desimone, 1990). If monkeys do not have another area in which cells show overt color opponency analogous to V-8 in humans (Hadjikhani et al, 1998), then V-1 and V-2 would be left with the task of subserving temporal color contrast. This would make V-1 the likely site of the

---

[1] Here I use the term Double-Opponent to describe the receptive fields of all cortical color cells that show overt opponency and have large receptive field centers. This includes cells having very weak surrounds (that may have been called Type II cells), cells having very strong surrounds, and cells having coarsely oriented Double-Opponent sub-fields.

generation of color percepts. Color percepts could not arise before V-1 because Double-Opponent cells, which are necessary for spatial color contrast, do not arise until then; and color percepts could not arise after V-1 because the overt cone-opponency necessary for temporal color contrast would be lost after V-1. If this is true, perhaps later "color" areas (like V-4) are involved in binding color and other stimulus attributes, like form (Schein and Desimone, 1990).

To my knowledge, no one has studied the responses of geniculate or cortical color cells using similar color-contrast and opposite color-contrast stimuli sequences. Others have examined the temporal and chromatic aspects of color cells in primary visual cortex (Cottaris and DeValois, 1998), but these studies involved determining the color preferences of cells using reverse correlation to full-field pseudo-random color sequences. Most of their color-time maps are consistent with the findings presented here — though it is difficult to interpret their maps because they were generated using full-field stimuli. Color cells have spatially structured receptive fields, and the surrounds have opposite chromatic tuning to the centers (Michael, 1978; Livingstone and Hubel; Conway, 2001). Thus full-field stimuli would have been problematic because they would have confounded the responses of center and surround. For example, some of their color-time maps (Cottaris and DeValois, 1998) imply that the best stimulus would be one whose color gradually shifted through the colors of the rainbow (over ~100ms) (Cottaris and DeValois, 1998, their Figure 1d). It is difficult to imagine what benefit such cells confer on their owners. Instead it seems more likely that color cells in V-1 are sensitive to sharp spatial and chromatic boundaries — the sort of cue that we use to detect and discriminate objects — and that these cues were confounded by using full-field stimuli.

Finally, I found that the temporal responses of V-1 color cells sum linearly (Figure 3.7). This is satisfying because most models of color vision require this (Wyszecki and Stiles, 1982, pg. 118).

3§4.2 Afterimages and the spatial variable

As discussed in Chapters 1 and 2, color perception is not equated to wavelength discrimination. This is demonstrated by the fact that the color we assign a given reflected light is not only affected by the wavelength composition of light reflected from surrounding regions but is also affected by chromatic adaptation (see above). As discussed above, chromatic adaptation underlies temporal color contrast (Figure 3.1) — a feature of our visual systems that could be useful in the wild. Consider a monkey viewing a wall of green leaves in which there is imbedded a small red fruit. As the monkey shifts his gaze over the scene the chromatic contrast of the red fruit is heightened because of the afterimage generated by the green leaves.

But under normal circumstances we do not see color afterimages. For example, after I look at a blue chair, the red magazine holder bears no impression of the blue chair superimposed on it. Goethe maintained that color afterimages "occur oftener than we are aware in ordinary life" and that it is "an attentive observer [who] sees these appearances everywhere, while, on the other hand, the uninstructed, . . . consider them as temporary visual defects, sometimes even as symptoms of disorders in the eye" (Goethe, 1967, pg. 21-22).

Perhaps with Goethe's greater attention came a more stable state of fixation. Maintaining unusually steady fixation results in "unusual phenomena . . . [where] sharply defined negative after-images of objects develop, which coincide with the objects as long as the gaze is held steady" (Helmholtz, 1909, pg. 266). Helmholtz thought that we did not normally see afterimages because "under ordinary circumstances, we are accustomed to let our eyes roam slowly about over the visual field continuously, so that the point of fixation glides from one part of the observed object to another. . . This wandering of the gaze serves to keep up on all parts of the retina a continual alternation between stronger and weaker stimulation, and between different colors, and is evidently of great significance for the normality and efficiency of the visual mechanism" (Helmholtz, 1909, pg. 266).

Neither Goethe's nor Helmholtz's explanation for the absence of afterimages under normal circumstances is wholly satisfying. Afterimages are probably continually present in the retina, and we probably do not see them because they are suppressed by more salient visual cues, like edges (Daw, 1962). Indeed, an afterimage will "reappear whenever the eye goes back to the original point of fixation" (Daw, 1962) suggesting that afterimages are inhibited by conflicting visual cues and are not destroyed.

A similar logic may help explain the curious observation that the afterimage of an induced color is much stronger than the induced color itself (Figure 3.9A). A gray disc surrounded by a large red annulus does have a greenish tinge (though it is very weak unless the annulus encompasses most of the visual field). But the afterimage produced by this image is a striking red disc surrounded by a green annulus (Figure 3.9A). Josef Albers constructed some similar illusions, consisting of panels of yellow dots. The average observer does not report any induced color between the dots, yet the afterimage of these spaces is a vibrant yellow (Figure 3.9B). I asked whether it was possible if the induced colors (of the gray spot and the white diamonds) are suppressed by the edges of the image, in the same way afterimages are suppressed by edges.

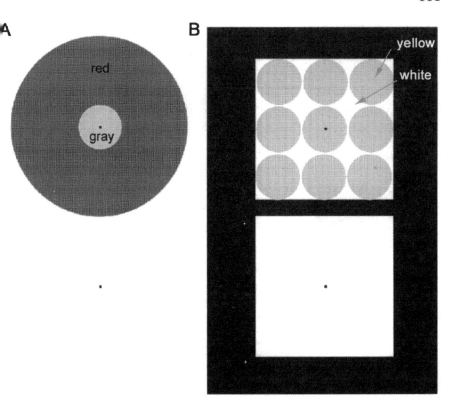

Figure 3.9. The spatial context of an image shapes the afterimage. **A.** The afterimage of the gray disc is a red spot. **B.** The afterimage of the white spaces between the yellow circles is yellow (From Albers, 1963, plate VIII-2).

We tested this in vision group meeting in the Department of Neurobiology at Harvard one Friday afternoon, by blurring the edges between a gray disc and a surrounding colored annulus. To our surprise the induced color of the gray disc was much stronger! Our perception of the color of the gray disc seems to take into account both the induced color and other visual cues, like the sharp edges of the disc and the luminance of the surrounding annulus. These other cues work to suppress the induced color. But in the afterimage condition, where one views a uniformly colored field, these cues are absent, leaving the afterimage of the induced color unsuppressed.

If Double-Opponent cells are mediating both spatial and temporal color contrast, it should not come as a surprise that the spatial configuration of an image affects both our perceptions of the color of a given surface within the image and our perceptions of the afterimage. But as Daw (1962) showed, spatial information seems to suppress color afterimages that are not in geometric register. One might expect Double-Opponent cells to reflect this. Indeed there is some evidence that they do: Double-Opponent cells respond preferentially to the edges of wide stimuli (Figure 3.4) but this is at the cost of the biphasic temporal response. With wide stimuli, the misalignment of the stimulus and the receptive field suppresses the off-excitation (to M-plus in Figure 3.4) and off-suppression (to L-plus in Figure 3.4), disabling the cell from encoding temporal color contrast.

In summary, Double-Opponent cells in V-1 are well equipped to serve both temporal and spatial color-contrast, though they respond more robustly to the spatial configuration of a stimulus than they do to the temporal configuration of it. This is consistent with the observation that afterimages are not normally seen. The lack of registration between afterimages and subsequent images projected on the retina under most circumstances suppresses our perceptions of afterimages, yet our perception of spatial color contrast is not affected.

Chapter 4

Segregated processing streams
in primary visual cortex

But no physiologist who calmly considers the question . . . can long resist the conviction that different parts of the cerebrum subserve different kinds of mental action. Localization of function is the law of all organization whatever: separateness of duty is universally accompanied with separateness of structure: and it would be marvelous were an exception to exist in the cerebral hemispheres. Let it be granted that the cerebral hemispheres are the seat of higher psychical activities; let it be granted that among these higher psychical activities there are distinctions of kind . . . more or less distinct parts of the cerebral hemispheres . . . It is proved experimentally, that every bundle of nerve fibers and every ganglion, has a special duty; and that each part of every bundle of nerve fibers and every such ganglion, has a duty still more special. Can it be, then, that in the great hemispherical ganglia alone, this specialization of duty does not hold?

— Spencer, *Principles of Psychology* (1855)

We take for granted the efficacy with which our visual systems process visual stimuli, but the apparent cohesion of our visual perception can be disrupted by clever visual illusions. For example, if a small circular disc is colored so that its brightness (but not its hue) is the same as the area surrounding the disc, it becomes very difficult to describe accurately any movement of the disc (Gregory, 1977). In fact, even if the disc is sitting still, we have the peculiar sensation that it is unstable, or shaking. Such "equiluminant" illusions show that separate aspects of our visual perception are constructed from separate physical cues: color perception obviously depends on the relative spectral distribution of reflected light (between the object and its surround), and motion perception depends on the relative luminance. But that we can selectively cripple motion perception without affecting color perception, as in the example just described, suggests that these different aspects of visual perception are mediated by different sets of neurons.

The general idea that the brain processes different aspects of the perception separately and simultaneously is not a new one, though it may have been rooted in superstition. Phrenology, for example, had little scientific basis — although accounts of specific deficits associated with well-defined cortical lesions can be found in the earliest medical records (Breasted, 1930). In fact, though Gall (1758-1828) is faulted with assigning different aspects of personality to different bumps on the skull, he did establish that careful anatomy is the basis for understanding brain function (he was the first to recognize that gray matter and white matter probably subserve different functions). "If ever there were to be a phrenological science, it would be the phrenology of convolutions, not bumps", said Broca (1824-1880), who was careful to distance himself from Gall while acknowledging functional localization as an important concept.

Gall's impact was wide reaching; of particular interest was his influence on the Italian physician Bartolomeo Panizza (1785-1867), who went on to use infarct patients to assign visual function to the occipital pole. It is of course now established that on some gross level, different regions of the brain subserve different functions: the motor cortex is localized to the central sulcus, the visual cortex to the occipital pole and so on. In more recent history, Mountcastle, working in the motor cortex, and Hubel and Wiesel, in the visual cortex, showed that the brain is subdivided even further into columns, of the order of millimeters: groups of cells that respond to similar stimulus attributes are clustered. In the visual cortex, for example, cells responsive to a particular stimulus orientation are grouped together (Hubel and Wiesel, 1962), as are cells responsive to stimulation of one eye (so-called ocular dominance columns, LeVay et al, 1975). But are different neurons responsible for form, color and motion perception segregated? Are there even different neurons for form, color and motion perception?

Segregated processing streams in the visual system

All classes of retinal ganglion cells sample the same retinal cone mosaic, yet different classes respond to different patterns of cone activity. The midget cells (i.e. Type I cells, Figure 1.6A), for example, have tiny receptive field centers: a given midget cell may have a center fed by a single cone (Polyak, 1957). The parasol cells, on the other hand, sample many cones over a relatively large region of retina (these different sampling abilities are reflected in the dendritic fields of these neurons, see Figure 4.2 below). Some have argued that these different classes of retinal ganglion cells represent the physiological substrate for separate aspects of visual perception: the midget cells subserving "form" and the parasol cells subserving "motion"(Livingstone and Hubel, 1988).

We know that visual perception is segregated in some form, at some level; the visual illusion described above (the isoluminant disc) provides one piece of evidence and the selective loss of color or motion perception in some stroke patients provides a second (Damasio et al., 1980; Rizzo et al., 1992; Sachs, 1995). But are these separate aspects of visual perception represented by separate channels as early as the retina? It is plausible, given that in monkeys specific lesions localized to specific layers of the lateral geniculate nucleus result in specific perceptual deficits, (Schiller et al., 1990). But how these perceptual deficits are related to cortical processing in extrastriate areas is unclear: unlike the perceptual deficits associated with human stroke patients, lesions in extrastriate areas of monkeys rarely produce selective deficits (Schiller,

1993); this suggests that either the lesions were not complete and therefore did not completely abolish the function of a given cortical area, or that processing of different attributes of the visual world are distributed among several extrastriate areas so that ablating any single cortical area has little effect on perception.

The question of the correlation between different subcortical "streams" (e.g. parvocellular, magnocellular and koniocellular layers of the lateral geniculate nucleus, see Figure 4.2) and different aspects of visual perception (form, motion and color) has generated much debate. Those who object to segregated processing streams prefer the notion that neurons throughout the visual pathway "multiplex" multiple aspects of the visual world. They find support for this with the observation that under perceptual conditions where motion perception *appears* to be compromised (like equiluminance) neurons in the lateral geniculate nucleus taken to underlie motion perception (the magnocellular cells) are still responsive (Logothetis et al, 1990). Instead of segregated streams, they argue that single cells subsequent to the retinal ganglion cells encode information about the form, color *and* motion of objects (e.g. see Lennie, 2000). From a perceptual standpoint, this argument has the merit that it correlates with the apparent cohesion of the various attributes associated with objects; for example, we perceive the "red" as "bound" to the red apple.

No one disagrees that discrete classes of retinal ganglion cells exist; nor do they dispute that different ganglion cell classes project to different layers of the lateral geniculate nucleus. But what this segregation means for perception is hotly contested. Moreover, in primary visual cortex, even the segregation of different classes of neurons (let alone the question of their correlation to perception!) is disputed (Lennie et al, 1990; Leventhal et al, 1995); in V-1 one finds orientation-selective cells, for example, that also seem to encode color

(Hubel and Wiesel, 1968; Livingstone and Hubel, 1984; Lennie et al, 1990).

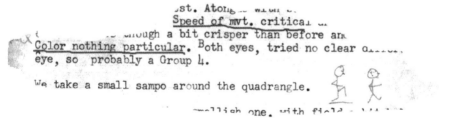

Figure 4.1. A fragment from the notebook of David Hubel and Torsten Wiesel, July 5, 1966, shortly before 9:15 pm, documenting a neuron that showed motion selectivity but not color selectivity.

Orientation-selective cells do not receive a stereotyped ratio of input from the three cone classes (Lennie et al, 1990). Rather, the relative contributions made by each cone class to a given cortical cell varies, and this results in different "color" selectivity for different neurons. Does this difference imply that these cells are encoding color information? It is difficult to say; but it would seem that it need not: why would evolution have gone out of its way to demand that each orientation-selective cell receive exactly the same ratio of cone inputs? But this line of reasoning is difficult, for obvious reasons: who gets to decide what represents biologically relevant variations in cone inputs?

Instead of relying on single cells to multiplex responses to multiple stimulus attributes (which extrastriate cortex would presumably have to "demultiplex") it seems simpler to contend that different aspects of the visual world are encoded by distinctly different populations of cells within V-1, each population responsible for one (or a few) visual attributes. This is not unreasonable: one of the earliest observations about neurons in primary visual cortex is that beyond subtle variations in color selectivity, some neurons show overt cone-opponency (Hubel and Wiesel, 1968; Michael, 1974), while others show selectivity for stimulus motion (Figure 4.1, Hubel and Wiesel, 1962). It would seem these cells are responsible for color and motion perception, leaving the majority of remaining cells, most of which tend to be strongly selective for the orientation of a stimulus, responsible for form perception (Livingstone and Hubel, 1987). That many of these orientation-selective cells also display some selectivity for the wavelength of the stimulus may simply reflect the fact that these cells are sampling a relatively few number of cones from a patchy cone mosaic. . . thus the "color"-selectivity of these cells may actually be a red herring in terms of color-coding *per se*. A similar argument has been used to rationalize the color-coding properties of Type I cells: if Type I cells sample a single cone in the center of their receptive fields then by default they will display some level of color selectivity (Lennie, 1980; Calkins and Sterling, 1999). [It does, however, seem unlikely that Type I cells achieve their color-selectivity only as a byproduct of high spatial resolution because Type I cells in the peripheral retina (whose receptive field centers are fed by several cones) also show cone opponency (Martin et al, 2001).]

The question of segregated processing streams in primary visual cortex would seem to have been resolved with the discovery that color-opponent neurons are localized to metabolically distinct regions dotted throughout primary visual cortex (the cytochrome oxidase blobs, Livingstone and Hubel, 1984; see Figure 4.2). But this claim, and even the claim that single cells in primary visual cortex are specialized for different functions, has been disputed (Lennie et. al., 1990; Leventhal et al., 1995). In this book, I sought to investigate the spatial and temporal structure of the receptive fields of cells in primary visual cortex, an issue I could readdress using the invention of eye-position corrected reverse-correlation (Livingstone et al., 1996). This technique enables one to make high-resolution

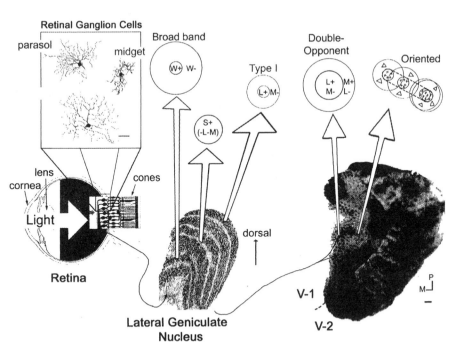

Figure 4.2. A summary of the visual system. Light enters the eye (at left) and is focused on the retina by the cornea and lens. The three classes of cones respond to the light. Different retinal ganglion cells (inset, adapted from Dacey and Lee (1994) *Nature* 367:731; scale bar 50 μm) sample the cone mosaic and provide input to the lateral geniculate nucleus (LGN). The cells of the LGN, here stained with Nissl substance, are comprised of six well-defined layers: four dorsal (or parvocellular) layers and two more darkly staining ventral (or magnocellular) layers. Each dot is a single cell, ~10 μm in diameter. The parvocellular layers contain Type I cells; the receptive field of an L-on-center/M-off-surround Type I cell is given. The magnocellular layers contain the Broadband cells. Between the darkly staining parvocellular and magnocellular layers are the intercalated (or koniocellular) layers. Type II cells reside in these layers. Broadband cells, Type II cells and Type I cells are the LGN targets of Parasol, Bistratified and Midget ganglion cells, respectively. (*continued*)

(*Figure 4.2 continued*) Neurons in the LGN send their axons to the primary visual cortex (V-1): magnocellular cells project to layer 4Cβ, parvocellular cells to 4Cβ and 4A (Hubel and Wiesel, 1972) and koniocellular cells to the cytochrome oxidase blobs at the base of 3B (Livingstone and Hubel, 1982). In this figure, V-1 is represented by a tangential section of flattened squirrel monkey cortex that has been stained with the metabolic enzyme cytochrome oxidase (the section is oriented with 'M' pointing towards the midline and 'P', posterior; scale bar is 1 mm). Cytochrome oxidase (CO) staining clearly demarcates the border between V-1 and the second visual area, V-2, and reveals CO blobs in V-1 and the CO stripes in V-2. Color information is processed by the Double-Opponent cells, which probably reside in the V-1 blobs and send their axons to the thin CO stripes of V-2 (arrows). Between the blobs are cells that are sensitive to the orientation of a visual stimulus.

receptive field maps while recording in alert animals. The use of alert monkeys has enormous advantages: one uses far fewer animals (than used in acute experiments) and the physiological state of the animal is closer to that during natural viewing conditions; moreover, the cells tend to be more responsive (Snodderly and Gur, 1995). In addition, it has always been a worry that select groups of neurons in V-1 are differently affected by anesthetics — one might worry, for example, that neurons in the cytochrome oxidase blobs might be more sensitive because of their higher metabolic state. If color cells do in fact reside in cytochrome oxidase blobs, such sensitivity might explain why color cells have been difficult to document in experiments done in more recent experiments on anesthetized animals.

The receptive fields of the thousand or so cells in V-1 that I studied support the notion that different neurons encode different stimulus attributes (Figure 3.2, for example). I found that a relatively small class of cells (10%) are strongly cone-opponent; many of them also have Double-Opponent receptive fields. These cells seem well equipped to encode spatial (Chapter 2) and temporal (Chapter 3) color contrast. One might have thought that this number of cells would be too few to encode the millions of colors which we can perceive, but computational studies have shown that this number would be sufficient (Buchsbaum and Gottschalk, 1983, 1984). Thus the visual system seems to work in a way analogous to a color monitor: monitors use less than 10% of the available bandwidth to represent color and the remaining 90% to represent luminance. It is perhaps not surprising that most of our cortical cells are devoted to luminance and form information (and not color) given that black-and-white pictures relay most of the information about a scene.

Double-Opponent cells likely reside in the cytochrome oxidase blobs of V-1 (Figure 4.2; Livingstone and Hubel, 1984; but see Lennie et al, 1990 and Leventhal et al, 1995). Perhaps not surprisingly, clustering of cells is a common feature in mammalian visual systems: the visual cortex of cats, for example, also contains cytochrome oxidase blobs (Dyck and Cynader, 1993; Murphy et al, 1995) and the cells in these blobs project selectively to distinct extrastriate visual areas (that likely process information about surfaces, of which color is one cue; Conway et al, 2000) . In monkeys, color signals carried by the blobs of V-1 are relayed to V-2 (the second visual area) and then to higher visual areas. Like V-1, V-2 displays an interesting pattern of staining for the enzyme cytochrome oxidase; unlike V-1, the staining manifests itself not as blobs but rather as dark thick and thin stripes, and faint inter-stripes separating these (Figure 4.2, bottom right). Cells residing in the blobs of V-1 specifically send their axons to the thin stripes of V-2 (Livingstone and Hubel, 1984). Not surprisingly, cells in the thin stripes are more likely to be color selective than cells in the thick stripes. The color cells in the V-2 thin stripes respond best to colored spots, but they do so over a larger area of visual space (Hubel and Livingstone, 1985, 1987). Importantly, they are not responsive to a large field of color that encompasses the entire region over which small spots are effective. Such "complex" color cells may be useful in identifying a color boundary present anywhere within a large area.

The V-2 thin stripes project to V-4 (the V-2 thick stripes, on the other hand, project to MT, an area specialized for analysing stimulus motion). Many (though not all) V-4 neurons respond to some wavelengths better than they respond to other wavelengths (Zeki, 1983), but their receptive fields are distinguished from those of color cells in V-1 in at least three ways. First, V-4 receptive fields are much larger, encompassing a larger part of the visual field. Second, V-4 neurons generally do not show overt opponency between cone classes. And third, many V-4 neuron receptive fields have a more extensive surround that antagonizes stimulation of the center. These surrounds are probably involved in further elaborating color constancy. One prominent color physiologist, Semir Zeki, claims that neurons in V-4 are the ones responsible for generating illuminant-independent (i.e. color-constant) color percepts (Zeki, 1983a,b).

In humans, an area (V-8) situated in the inferotemporal cortex is specialized for computing color percepts (Hadjikhani et al, 1998). In fact, certain stroke victims who have damage to this region show a profound loss of color perception. Remarkably this acquired achromatopsia does not interfere with their perception of form and motion — further suggesting that color is processed separately from other visual attributes. Oliver Sachs described one such stroke patient who was an artist (Sachs, 1995). After the patient had the stroke and lost his color vision, he made peculiar color choices in his paintings. But he was perfectly able to represent luminance and shape because the areas of his brain dedicated to processing those aspects of the visual world were unaffected by the stroke. Moreover, he did not have any loss of motion perception.

Area V-8 in humans might correspond to area V-4 in monkeys (Zeki et al, 1998; but see Tootell and Hadjikhani, 1998), though comparisons between species are difficult beyond V-1 and V-2. Neurons in V-4 of monkeys are color biased; but they are also selective for other attributes of the stimulus, such as the stimulus orientation (Schein and Desimone, 1990), suggesting that V-4 is not simply a color area. To complicate the matter further, unlike lesions of V-8 in humans, partial lesions of V-4 in monkeys have little effect on tasks requiring color vision. Perhaps V-4 is involved in piecing together form and color information; conversely, V-4 may actually be a complex of areas, one devoted to form processing and another to color processing.

An observer would say that the color and the shape of an object are inextricably linked. If color and form are processed separately by the cortex, how do they then become bound? How would this binding be manifest in the brain? The binding might simply be found in the correlated activity of the two pathways: an orange ball would produce separate sensations of "round" in the form pathway and "orange" in the color pathway, but the sensations would be elicited simultaneously. The reliability of the simultaneous activation of the two pathways, possibly reinforced by neural connections joining separate areas, could be enough to bind the "round" ball with its "orange" color.

Parallel processing is an efficient means of computing information because it is fast: multiple aspects of a scene can be processed simultaneously. But in providing this as a model for visual processing we risk making premature conclusions. In this spirit, it is worth noting that there is much about color processing, even in the early stages, about which we are ignorant. There are cells in the superficial layers of V-1 (not confined to blobs) that are both selective for stimulus orientation and also more responsive to some wavelengths than others (Livingstone and Hubel, 1984; Lennie et al, 1990). What contribution, if any, do such "oriented color" cells make to color perception? What are the LGN inputs to red-green Double-Opponent cells? Are there retinal and/or geniculate red-green Type II cells? Do red-green Double-Opponent cells get S cone input? We have more red-green Double-Opponent cells than blue-yellow Double-Opponent cells —why? What is the role of cortical Type I cells in color perception? Which cells underlie the black-white color axis? How is the activity of the three chromatic axes in V-1 integrated to effect the perception of specific hues? And, how do colors bring about emotional responses?

In the words of Elio Raviola, "one must be careful to distinguish a fact from a hypothesis": we do not *know* if only one class of cells is solely responsible for one aspect of visual perception. It almost seems like an absurd proposal (why would evolution have restricted itself to using only one class of cells if there are advantages to using several?); but to suppose that all neurons are responsible for every aspect of visual perception seems equally absurd. Instead, some balance is likely: distinct classes of cells exist (e.g., color cells, direction-selective cells, orientation-selective cells) that provide the main input to separate extrastriate areas subserving different aspects of visual perception (color, motion, and form; e.g. Shipp and Zeki, 1985), but there is some cross-talk between different classes of cells and separate extrastriate areas which possibly helps reinforce (and perhaps "bind") the perceptions subserved by different streams. Thus one would not expect a *complete* loss of color at equiluminance, or a *complete* loss of motion perception after an MT lesion; this represents the remarkable success of evolution in making visual perception both cohesive and robust.

REFERENCES

Adrian, E. D., and Matthews, R. (1926). The action of light on the eye. Part 1. The discharge of impulses in the optic nerve and its relation to the elecric changes in the retina. *J. Physiol.* **63**, 378-414.

Albers, J. (1963). Interaction of Color. New Haven, London, Yale University Press.

Anderson, J. C., Martin, K. A., and Whitteridge, D. (1993). Form, function, and intracortical projections of neurons in the striate cortex of the monkey *Macacus nemestrinus*. *Cerebral Cortex* **3**, 412-20.

Baylor, D. A., Nunn, B. J., and Schnapf, J. L. (1987). Spectral sensitivity of cones of the monkey *Macaca fascicularis*. *J. Physiol.* **390**, 145-60.

Bowmaker, J. K., Dartnall, H. J. and Mollon, J. D. (1980). Microspectrophotometric demonstration of four classes of photoreceptor in an old world primate, *Macaca fascicularis*. *J. Physiol.* **298**, 131-43.

Breasted, J. H. (1930). The Edwin Smith Surgical Papyrus. University of Chicago Press, Chicago, Illinois.

Brindley, C. S. (1957). Two theorems of colour vision. *Q. J. Exp. Psychol.* **9**, 101.

Brindley, G. S. (1962). Two new properties of foveal after-images and a photochemical hypothesis to explain them. *J. Physiol* **164**, 168-179.

Brown, P. K., and Wald, G. (1963). Visual pigments in human and monkey retinas. *Nature* **200**, 37-43.

Brown, S. C. e. (1970). Collected works of Count Rumford. Harvard University Press, Cambridge, Massachusetts.

Buchsbaum, G. and Gottschalk, A. (1983). Trichromacy, opponent colours coding and optimum colour information transmission in the retina. *Proc. R. Soc. Lond. Series B. Biol. Sci.* **220**, 89-113.

Buchsbaum, G. and Gottschalk, A. (1984). Chromaticity coordinates of frequency-limited functions. *J. Opt. Soc. Am.A-Optics and Image Science* 1, 885-7.

Burr, D. C., Fiorentini, A., and Morrone, M. C. (1987). Electrophysiological correlates of positive and negative afterimages. *Vision Res.* 27, 201-7.

Calkins, D. J., and Sterling, P. (1999). Evidence that circuits for spatial and color vision segregate at the first retinal synapse. *Neuron* 24, 313-21.

Calkins, D. J., Tsukamoto, Y., and Sterling, P. (1998). Microcircuitry and mosaic of a blue-yellow ganglion cell in the primate retina. *J. Neurosci.* 18(9): 3373-85.

Chichilnisky, E. J., and Baylor, D. A. (1999). Receptive-field microstructure of blue-yellow ganglion cells in primate retina. *Nat. Neurosci.* 2, 889-893.

Conway, B. R. (2000). Spatial organization of cone contributions to color cells in alert macaque primary visual cortex. *Soc. Neurosci. Abstr.* 26:54.14.

Conway, B. R. (2001). Spatial structure of cone inputs to color cells in alert macaque primary visual cortex (V-1). *J. Neurosci.* 21(8), 2768-2783.

Conway, B. R., Hubel, D. H., and Livingstone, M.S. (2002). Color contrast in macaque V-1. *Cerebral Cortex*, in press

Conway, B., Boyd, J. D., Stewart, T. H. and Matsubara, J. A. (2000). The projection from V1 to extrastriate area 21a: a second patchy efferent pathway colocalizes with the CO blob columns in cat visual cortex. *Cerebral Cortex* 10, 149-59.

Cottaris, N. P., Elfar, S.D., and De Valois R.L. (2000). Strong S-con inputs to macaque V-1 simple cell spatio-temporal chromatic receptive fields corrected for axial chromatic aberration. *Soc. Neurosci. Abstr.* 26:54.11.

Courtney, S. M., Finkel, L. H., and Buchsbaum, G. (1995). Network simulations of retinal and cortical contributions to color constancy. *Vision Res.* 35(3): 413-34.

Cottaris, N. P., and De Valois, R. L. (1998). Temporal dynamics of chromatic tuning in macaque primary visual cortex. *Nature* **395**, 896-900.

Dacey, D. M., and Lee, B. B. (1994). The 'blue-on' opponent pathway in primate retina originates from a distinct bistratified ganglion cell type. *Nature* **367**, 731-5.

Damasio, A., T. Yamada, Damasio, H., Corbett, J., and McKee, J. (1980).Central achromatopsia: behavioral, anatomic, and physiologic aspects. *Neurology* **30**(10): 1064-71.

Dartnall, H. J., Bowmaker, J. K. and Mollon, J. D. (1983). Human visual pigments: microspectrophotometric results from the eyes of seven persons. *Proc. R. Soc. Lond. Series B. Biol. Sci.* **220**, 115-30.

Daw, N. (1968). Goldfish retina: organization for simultaneous color contrast. *Science* **158**, 942-944.

Daw, N. W. (1962). Why after-images are not seen in normal circumstances. *Nature* **196**.

Daw, N. W. (1972). Color-coded cells in goldfish, cat, and rhesus monkey. *Investigative Ophthalmology* **11**, 411-7.

De Valois, K. K., and Switkes, E. (1983). Simultaneous masking interactions between chromatic and luminance gratings. *J. Opt. Soc. Am.* **73**, 11-8.

De Valois, R. L., Smith, C.J., Kitai, S.T., and Karoly, A.J. (1958). Response of single cells in monkey lateral genculate nucleus to monochromatic light. *Science* **127**, 238-239.

De Valois, R. L., Morgan, H. C., Polson, M. C., Mead, W. R., and Hull, E. M. (1974). Psychophysical studies of monkey vision. I. Macaque luminosity and color vision tests. *Vision Res.* **14**, 53-67.

Derrington, A. M., Krauskopf, J., and Lennie, P. (1984). Chromatic mechanisms in lateral geniculate nucleus of macaque. *J. Physiol.* **357**, 241-65.

Diamond, I. T., Conley, M., Itoh, K., and Fitzpatrick, D. (1985). Laminar organization of geniculocortical projections in Galago senegalensis and Aotus trivirgatus. *J. Comp. Neurol.* **242**, 584-610.

Donner, K. O., and Rushton, W. A. H. (1959). Retinal stimulation by light substitution. *J. Physiol.* **149**, 288-302.

Dow, B. M. (1974). Functional classes of cells and their laminar distribution in monkey visual cortex. *J. Neurophysiol.* **37**, 927-946.

Dow, B. M., and Gouras, P. (1973). Color and spatial specificity of single units in Rhesus monkey foveal striate cortex. *J. Neurophysiol.* **36**, 79-100.

Dufort, P. A., and Lumsden, C. J. (1991). Color categorization and color constancy in a neural network model of V4. *Biol. Cybernetics* **65**, 293-303.

Dyck, R. H. and Cynader, M. S. (1993). An interdigitated columnar mosaic of cytochrome oxidase, zinc, and neurotransmitter-related molecules in cat and monkey visual cortex. *Proc. Natl. Acad. Sci. USA* **90**, 9066-9.

Estevez, O., and Spekreijse, H. (1982). The "silent substitution" method in visual research. *Vision Res.* **22**, 681-91.

Ferster, D. (1994). Linearity of synaptic interactions in the assembly of receptive fields in cat visual cortex. *Curr. Op. Neurobiol.* **4**, 563-568.

Fitzpatrick, D., Lund, J. S., and Blasdel, G. G. (1985). Intrinsic connections of macaque striate cortex: afferent and efferent connections of lamina 4C. *J. Neurosci.* **5**, 3329-3349.

Gegenfurtner, K. R., Kiper, D. C., and Fenstemaker, S. B. (1996). Processing of color, form, and motion in macaque area V2. *Vis. Neurosci.* **13**, 161-72.

Geisler, W. S. (1978). Adaptation, afterimages and cone saturation. *Vision Res.* **18**, 279-89.

Gerling, J., and Spillmann, L. (1987). Duration of visual afterimages on modulated backgrounds: postreceptoral processes. *Vision Res.* **27**, 521-7.

Goethe. (1967). Theory of Colors. Translated by Eastlake, C. L., Frank Cass and Co. Ltd, London (first published in 1810).

Gouras, P. (1970). Trichromatic mechanisms in single cortical neurons. *Science* **168**, 489-92.

Gouras, P., and Kruger, J. (1979). Responses of cells in foveal visual cortex of the monkey to pure color contrast. *J. Neurophys.* **42**(3): 850-60.

Granger, E. M., and Heurtley, J. C. (1973). Letters to the editor: Visual chromaticity-modulation transfer function. *J. Opt. Soc. Am.* **63**, 1173-4.

Gregory, R. L. (1977). Vision with equiluminant color contrast: 1. A projection technique and observations. *Perception* **6**: 113-119.

Hadjikhani, N., Liu, A. K., Dale, A. M., Cavanagh, P., and Tootell, R. B. (1998). Retinotopy and color sensitivity in human visual cortical area V8 [see comments]. *Nat. Neurosci.* **1**, 235-41.

Hartline, H. K. (1939). The response of single optic nerve fibers of the vertebrate eye to illumination of the retina. *Amer. J. Physiol.* **121**, 400-415.

Hartline, H. K. (1949). Inhibition of activity of visual receptors by illuminating nearby retinal areas in the Limulus eye. *Fed. Proc.* **8**, 69.

Hartline, H. K., Wagner, H.G., and Ratliff, F. (1956). Inhibition in the eye of the limulus. *J. Gen. Physiol.* **39**, 651-673.

Helmholtz, H., translated by E. Atkinson. (1873). Popular lectures on Scientific Subjects. Ny d Apleton and co.

Helmholtz, H. v. (1909). Hanbuch der Physiologischen Optik. J. P. Southall, ed., Dover Publications, New York, New York.

Hering, E., translated by Hurvich, L.M. and Jameson, D. (1964). Outlines of a theory of the light sense. Harvard University Press, Cambridge Massachusetts .

Hess, R. F., and Nordby, K. (1986). Spatial and temporal limits of vision in the achromat. *J. Physiol.* **371,** 365-85.

Hubel, D. H. (1957). Tungsten microelectrode for recording from single units. *Science* **125,** 549.

Hubel, D. H. and Livingstone, M. S. (1985). Complex-unoriented cells in a subregion of primate area 18. *Nature* **315,** 325-7.

Hubel, D. H., and Livingstone, M. S. (1987). Segregation of form, color, and stereopsis in primate area 18. *J. Neurosci.* **7,** 3378-415.

Hubel, D., and Livingstone, M. (1990). Color puzzles. *Cold Spring Harbor Symposia on Quantitative Biology* **55,** 643-9.

Hubel, D. H., and Livingstone, M. S. (1990). Color and contrast sensitivity in the lateral geniculate body and primary visual cortex of the macaque monkey. *J. Neurosci.* **10**(7): 2223-37

Hubel, D. H., and Wiesel, T. N. (1962). Receptive fields, binocular interaction and functional architecture in the cat's visual cortex. *J.Physiol.* **160,** 106-154.

Hubel, D. H., and Wiesel, T. N. (1968). Receptive fields and functional architecture of monkey striate cortex. *J.Physiol.* **195,** 215-243.

Hubel, D. H., and Wiesel, T. N. (1972). Laminar and columnar distribution of geniculo-cortical fibers in the macaque monkey. *J Comp Neurol* **146,** 421-50.

Hurvich, L. M. (1981). Color vision. Sinauer Associates Inc., Boston, Massachusetts.

Ingling, C. R., Jr. (1977). The spectral sensitivity of the opponent-color channels. *Vision Res.* **17,** 1083-9.

Itten, J., 1888-1967. (1966). The art of color; the subjective experience and objective rationale of color. Translated by Ernst van Haa. Reinhol, New York, New York.

Johnson, E. N., Hawken, M. J., Shapley, R. (2001). The spatial transformation of color in the primary visual cortex of the macaque monkey. *Nat. Neurosci.* **4**, 409-416.

Judd, D. B. (1949). Response functions for types of vision according to the Muller theory. *J. Res. Nat. Bur. Standards (Washington DC)* **42**.

Judd, D. B. (1951). Basic correlates of the visual stimulus. in Handbook of Experimental Psychology, Chapter 22, S.S. Stevens, Editors. Wiley, New York, New York.

Kelly, J. P., and Van Essen, D. C. (1974). Cell structure and function in the visual cortex of the cat. *J.Physiol.(Lond)* **238**, 515-547.

Kiper, D. C., Fenstemaker, S. B., and Gegenfurtner, K. R. (1997). Chromatic properties of neurons in macaque area V2. *Vis. Neurosci.* **14**, 1061-72.

Kraft, J. M. and D. H. Brainard (1999). Mechanisms of color constancy under nearly natural viewing. *Proc. Nat. Acad. Sci. USA* **96**(1): 307-12.

Krauskopf, J., Williams, D. R., and Heeley, D. W. (1982). Cardinal directions of color space. *Vision Res.* **22**, 1123-31.

Kuffler, S. W. (1953). Discharge patterns and functional organization of mammalian retina. *J. Neurophysiol* **16**.

Land, E. (1959a). Experiments in color vision. *Scientific American* **May,** 2-14.

Land, E. H. (1959b). Color vision and the natural image. *Proc. Natl. Acad. Sci. U.S.A.* **45**, 115-129, 634-644.

Land, E. H. (1977). The retinex theory of color vision. *Scientific American* **237**, 108-28.

Land, E. H., Hubel, D.H., Livingstone, M.S., Perry, S.H., and Burns, M.M. (1983). Color-generating interactions across the corpus callosum. *Nature* **303**, 616-618.

Lennie, P. (1980). Parallel visual pathways: a review. *Vision Research* **20**, 561-94.

132

Lennie, P. and D'Zmura, M. (1988). Mechanisms of color vision. *Critical Reviews in Neurobiology* **3**, 333-400.

Lennie, P. (1999). Color coding in the cortex. In Color Vision (Eds. Gegenfurtner, K.R., Sharpe, L.T.), Cambridge University Press, Cambridge, UK.

Lennie, P. (2000). Color Vision. In <u>Principles of Neural Science</u> (4th Edition, Eds. Kandel, E.R., Schwartz, J.H., and Jessel, T. M). McGraw Hill Companies, Inc., New York, New York.

Lennie, P., Krauskopf, J., and Sclar, G. (1990). Chromatic mechanisms in striate cortex of macaque. *J. Neurosci.***10**, 649-69.

LeVay, S., Hubel, D. H. and Wiesel, T. N. (1975). The pattern of ocular dominance columns in macaque visual cortex revealed by a reduced silver stain. *J. Comp. Neurol.***159**, 559-76.

Leventhal, A. G., Thompson, K. G., Liu, D., Zhou, Y., and Ault, S. J.(1995). Concomitant sensitivity to orientation, direction, and color of cells in layers 2, 3, and 4 of monkey striate cortex. *J. Neurosci.***15**: 1808-18.

Liebmann, S. (1926). Uber das Verhalten farbiger Formen bei Helligkeitsgleichheit von Figur und Grund. *Psychol. Forsch.* **9**, 300-353.

Livingstone, M. S. (2002). <u>Vision and Art: the Biology of Seeing</u>. Harry N. Abrams Inc., New York, New York.

Livingstone, M. S., Freeman, D. C., and Hubel, D. H. (1996). Visual responses in V1 of freely viewing monkeys. *Cold Spring Harbor Symposia on Quantitative Biology* **61**, 27-37.

Livingstone, M. S., and Hubel, D. H. (1982). Thalamic inputs to cytochrome oxidase-rich regions in monkey visual cortex. *Proc. Natl. Acad. Sci. U.S.A.* **79**, 6098-6101.

Livingstone, M. S., and Hubel, D. H. (1984). Anatomy and physiology of a color system in the primate visual cortex. *J. Neurosci.* **4**, 309-356.

Livingstone, M. S., and Hubel, D. H. (1987). Psychophysical evidence for separate channels for the perception of form, color, movement, and depth. *J. .Neurosci.* **7**, 3416-3468.

Livingstone, M. and D. Hubel (1988). Segregation of form, color, movement, and depth: anatomy, physiology, and perception. *Science* **240**(4853): 740-9.

Livingstone, M. S., and Tsao, D. Y. (1999). Receptive fields of disparity-selective neurons in macaque striate cortex. *Nat. Neurosci.* **2**, 825-32.

Logothetis, N. K., Schiller, P. H. , Charles, E. R., and Hurlbert, A. C. (1990). Perceptual deficits and the activity of the color-opponent and broad-band pathways at isoluminance. *Science* **247**(4939): 214-7.

Macknik, S. L. and Livingstone, M. S. (1998). Neuronal correlates of visibility and invisibility in the primate visual system. *Nat. Neurosci.***1**, 144-9.

MacNichol, E. F. (1964). Retinal mechanisms of color vision. *Vision Res.* **4**(1): 119-33.

Marks, W. B., Dobelle, W.H., and MacNichol, E.F. Jr. (1964). Visual pigments of single primate cones. *Science* **143**, 1181--3.

Martin, P. R., Lee, B. B., White, A. J., Solomon, S. G. and Ruttiger, L. (2001). Chromatic sensitivity of ganglion cells in the peripheral primate retina. [see comments]. *Nature* **410**, 933-6.

Martin, P. R., White, A. J., Goodchild, A. K., Wilder, H. D., and Sefton, A. E. (1997). Evidence that blue-on cells are part of the third geniculocortical pathway in primates. *Eur. J. Neurosci.* **9**, 1536-41.

Maxwell, J. C. (1856). Theory of the perception of colors. *Transactions of the Royal Scottish Society of Arts* **4**, 394-400.

Michael, C. R. (1978). Color vision mechanisms in monkey striate cortex: dual-opponent cells with concentric receptive fields. *J. Neurophysiol.* **41**, 572-88.

Mollon, J. D. (1993). George Palmer (1740-1795). in <u>Dictionary of national biography: missing persons volume</u> Oxford university press pp. 509-510, Oxford, UK.

Movshon, J. A., Thompson, I. D. and Tolhurst, D. J. (1978). Receptive field organization of complex cells in the cat's striate cortex. *J. Physiol.* **283**, 79-99.

Mullen, K. T. (1985). The contrast sensitivity of human color vision to red-green and blue-yellow chromatic gratings. *J. Physiol.* **359**, 381-400.

Muller, G. E. (1930). Uber die Farbenempfindungen. *Z. Psychol., Erganzungsb.* **17 and 18**.

Murphy, K. M., Jones, D. G. and Van Sluyters, R. C. (1995). Cytochrome-oxidase blobs in cat primary visual cortex. *J. Neurosci.* **15**, 4196-208.

Nathans, J. (1999). The evolution and physiology of human color vision: insights from molecular genetic studies of visual pigments. *Neuron* **24**, 299-312.

Newton, I. (1978). <u>Isaac Newton's papers and letters on natural philosophy,</u> 2d ed (I.B. Cohen, R.E. Schofield, Eds). Harvard University Press, Cambridge, Massachusetts.

Newton, I. (1671). A Letter of Mr. Isaac Newton, Professor of the Mathematicks in the University of Cambridge; Containing His New Theory about Light and Colors: Sent by the Author to the Publisher from Cambridge, Febr. 6 1671/2; In Order to be Communicated to the R. Society. *Phil. Trans. R. Soc. Lond.* **6**, 3075-3087.

Ohzawa, I., DeAngelis, G. C., and Freeman, R. D. (1997). Encoding of binocular disparity by complex cells in the cat's visual cortex. *J. Neurophysiol.* **77**, 2879-909.

Poggio, G. F., Baker, F. H., Mansfield, R. J., Sillito, A., and Grigg, P. (1975). Spatial and chromatic properties of neurons subserving foveal and parafoveal vision in rhesus monkey. *Brain Res.* **100**, 25-59.

Polyak, S. e. b. K., H. (1957). <u>The vertebrate visual system</u>. The University of Chicago Press, Chicago, Illinois.

Reid, R. C., and Shapley, R. M. (1992). Spatial structure of cone inputs to receptive fields in primate lateral geniculate nucleus. *Nature* **356**, 716-718.

Rizzo, M., M. Nawrot, Blake, R., and Damasio, A. (1992). A human visual disorder resembling area V4 dysfunction in the monkey. *Neurology* **42**(6): 1175-80.

Rodieck. (1991). Which cells code for color? In From Pigments to Perception (Valberg, A and Lee, B.B., Eds.) Plenum Press, New York, New York.

Rodieck, R. W. (1998). The first steps in seeing. Sinauer, Sunderland Massachusetts.

Rodieck, R. W., and Watanabe, M. (1993). Survey of the Morphology of Macaque Retinal Ganglion Cells That Project to the Pretectum, Superior Colliculus, and Parvicellular Laminae of the Lateral Geniculate Nucleus. *J. Comp. Neurol.* **338**, 289-303.

Roorda, A. and Williams, D. R. (1999). The arrangement of the three cone classes in the living human eye [see comments]. *Nature* **397**, 520-2.

Rubin, J. M. and Richards, W. A. (1982). Color vision and image intensities: when are changes material? *Biol. Cybernetics* **45**, 215-26.

Rushton, W. A. (1972). Pigments and signals in colour vision. *J. Physiol.* **220**, 1P-P.

Rushton. (1975). Visual pigmenst and color blindness. *Scientific American* **March,** 64-74.

Rushton, W. A. H. (1958). Kinetics of cone pigments measured objectively in the living human fovea. *Annals of the New York Academy of Science* **74**, 291-304.

Rushton, W. A. H. (1961). Dark-adaptation and the regeneration of rhodopsin. *J. Physiol.* **156**, 166-78.

Rushton, W. A. H. (1962). Visual Pigments in Man. *Scientific American* **November.**

136

Sachs, O. (1995). An anthropologist on mars. Alfred A. Knopf, New York, New York.

Sandell, J. H., Gross, C. G., and Bornstein, M. H. (1979). Color categories in macaques. *J. Comp. Physiol. Psychol.* **93**, 626-35.

Schein, S. J., and Desimone, R. (1990). Spectral properties of V4 neurons in the macaque. *J. Neurosci.* **10**, 3369-89.

Schiller, P. H., Logothetis, N. K., and Charles, E. R. (1990). Role of the color-opponent and broad-band channels in vision. *Vis. Neurosci.* **5**, 321-346.

Schiller, P. H. (1993). The effects of V4 and middle temporal (MT) area lesions on visual performance in the rhesus monkey. *Vis. Neurosci.* **10**(4): 717-46.

Seidemann, E., Poirson, A. B., Wandell, B. A., and Newsome, W. T. (1999). Color signals in area MT of the macaque monkey. *Neuron* **24**, 911-7.

Shipp, S. and Zeki, S. (1985). Segregation of pathways leading from area V2 to areas V4 and V5 of macaque monkey visual cortex. *Nature* **315**(6017): 322-5

Smith, V. C., and Pokorny, J. (1972). Spectral sensitivity of color-blind observers and the cone photopigments. *Vision Res.* **12**, 2059-71.

Smith, V. C., and Pokorny, J. (1975). Spectral sensitivity of the foveal cone photopigments between 400 and 500 nm. *Vision Res.* **15**, 161-71.

Snodderly, D. M., and Gur, M. (1995). Organization of Striate Cortex of Alert, Trained Monkeys (Macaca fascicularis): Ongoing Activity, Stimulus Selectivity, and Widths of Receptive Field Activating Regions. *J. Neurophys.* **74**, 2100-2125.

Spillman, L., and Werner, J. S. (1990). Visual Perception: The Neurophysiological Foundations. Academic Press, San Diego, California.

Stiles, W. S. (1939). The directional sensitivity of the retina and the spectral sensitivities of the rods and cones. *Proc. Roy. Soc., B.* **127**, 64-105.

Stockman, A., and Sharpe, L. T. (2000). The spectral sensitivities of the middle- and long-wavelength-sensitive cones derived from measurements in observers of known genotype. *Vision Res.* **40**, 1711-1737.

Stockman, A., Sharpe, L. T., and Fach, C. (1999). The spectral sensitivity of the human short-wavelength sensitive cones derived from thresholds and color matches. *Vision Res.* **39**, 2901-27.

Svaetichin, G. (1956). Spectral response curves from single cones. *Acta Physiol. Scand. Suppl.* **39 suppl. 134**: 17-46.

Svaetichin, G., and MacNichol, E.F. (1958). Retinal mechanisms for chromatic and achromatic vision. *Ann. N.Y. Acad. Sci.* **74**, 385-404.

Tailor, D. R., Finkel, L. H., and Buchsbaum, G. (2000). Color-opponent receptive fields derived from independent component analysis of natural images. *Vision Res.* **40**(19): 2671-6.

Thibos, L. N., Bradley, A., Still, D. L., Zhang, X., and Howarth, P. A. (1990). Theory and measurement of ocular chromatic aberration. *Vision Res.* **30**, 33-49.

Tomita, T. (1957). A study on the origin of interretinal action potential of the cyprinid fish by means of pencil-type microelectrode. *Jap. J. Physiol.* **7**: 80-87.

Thomson, Sir Benjamin Count of Rumford (1794). An account of some experiments of colored shadows. *Philos. Trans. R. Soc. Lond.* **84**: 107-118.

Thorell L. G., De Valois R. L., Albrecht D. G. (1984) Spatial mapping of monkey V1 cells with pure color and luminance stimuli. *Vision Res.* **24** 751-69.

Tolhurst, D. J., and Dean, A. F. (1990). The effects of contrast on the linearity of spatial summation of simple cells in the cat's striate cortex. *Exp. Brain Res.* **79**, 582-8.

Tootell, R. B. H., and Hadjikhani, N. (1998). REPLY TO "Has a new color area been discovered". *Nat. Neurosci.* **1**(5): 335-3336.

138

Ts'o, D. Y., and Gilbert, C. D. (1988). The organization of chromatic and spatial interactions in the primate striate cortex. *J. Neurosci.* **8**, 1712-27.

Vautin, R. G., and Dow, B. M. (1985). Color cell groups in foveal striate cortex of the behaving macaque. *J. Neurophysiol.* **54**, 273-92.

Virsu, V., and Laurinen, P. (1977). Long-lasting afterimages caused by neural adaptation. *Vision Res.* **17**, 853-60.

Vos, J. J. and Walraven, P. L. (1971). On the derivation of the foveal receptor primaries. *Vision Res.* **11**, 799-818.

Wachtler, T., Lee, T. W., and Sejnowski, T. J .(2001). Chromatic structure of natural scenes. *J .Opt. Soc. Am., A, Optics, Image Science, & Vision* 18(1): 65-77.

Wandell, B. A. (1995). Foundations of Vision. Sinauer Associates Inc., Sunderland, Massachusetts.

Webster, M. A., and Mollon, J. D. (1991). Changes in color appearance following post-receptoral adaptation. *Nature* **349**, 235-8.

Wiesel, T. N., and Hubel, D. H. (1966). Spatial and chromatic interactions in the lateral geniculate body of the rhesus monkey. *J. Neurophys.* **29**, 1115-56.

Wyszecki, G., and Stiles, W. S. (1982). Color Science: concepts and methods, quantitative data and formulae. John Wiley and Sons, Inc., New York, New York.

Young, T. (1802). On the theory of light and colors. *Phil. Trans. R. Soc. Lond.* **92**, 12-48.

Zeki, S. (1983a). Color coding in the cerebral cortex: the responses of wavelength-selective and color-coded cells in monkey visual cortex to changes in wavelength composition. *Neuroscience* **9**, 767-81.

Zeki, S. (1983b). The relationship between wavelength and color studied in single cells of monkey striate cortex. *Progress in Brain Research* **58**, 219-27.

Zeki, S., D. J. McKeefry, Bartels, A., and Frackowiak, R. S.(1998). Has a new color area been discovered? *Nat. Neurosci.* 1(5): 335-6.

FURTHER READING

Gegenfurtner, K. R., and Sharpe, L. T. (Eds) (1999). Color Vision. Cambridge University Press, Cambridge, U.K.

Hubel, D. H. (1995). Eye, Brain and Vision. Scientific American Library, New York, New York.

Hurvich, L. M. (1981). Color vision. Sinauer Associates Inc., Sunderland, Massachusetts.

Livingstone, M. S. (2002) Vision and Art: the Biology of Seeing. Harry N. Abrams, Inc., New York, New York.

Wandell, B. A. (1995). Foundations of Vision. Sinauer Assoiciates Inc, Sunderland, Massachusetts.

Zeki, S. (1993). A vision of the brain. Blackwell Scientific Publications, Cambridge, Massachusetts.

INDEX *Authors*

INDEX *Terms*